T0202391

Data Science Ethics

Data Science Ethics

Concepts, Techniques and Cautionary Tales

DAVID MARTENS

OXFORD
UNIVERSITY PRESS

Great Clarendon Street, Oxford, OX2 6DP,
United Kingdom

Oxford University Press is a department of the University of Oxford.
It furthers the University's objective of excellence in research, scholarship,
and education by publishing worldwide. Oxford is a registered trade mark of
Oxford University Press in the UK and in certain other countries

Published in the United States of America by Oxford University Press
198 Madison Avenue, New York, NY 10016, United States of America

British Library Cataloguing in Publication Data

Data available

Library of Congress Control Number: 2021946685

ISBN 978–0–19–284726–3 (hbk)
ISBN 978–0–19–284727–0 (pbk)

DOI: 10.1093/oso/9780192847263.001.0001

Printed and bound by
CPI Group (UK) Ltd, Croydon, CR0 4YY

Cover image: inimalGraphic/Shutterstock.com.

Foreword

Can you confidently say what are the ethical pitfalls involved with implementing products and services based on artificial intelligence, machine learning, or other data-science technologies?

If you can't, this book is for you.

I met David more than a decade ago. He is a successful scholar, winning awards for his data science research, and a seasoned practitioner who has worked to build multiple data-science-based companies. What has impressed me most, though, is how great his students are. When one after another of a professor's students is great, you have to turn and take another look at the professor. David's combination of deep understanding of data science and deep practical experience gives his students-and readers-a healthy and realistic perspective on what is truly important in the world of data science.

This combination of scholarly expertise and practical experience has resulted in a book that fills an important gap.

Writings on data science and ethics generally fall into one of two categories. First we have scholarly articles, written for other scholars. Although I write them myself, I have a very difficult time recommending scholarly articles to my practically oriented students-seldom do they contain clear practical lessons and often they are not even accessible to practitioners. The other main category is writings in the popular press on the collision of data science and ethics. Unfortunately, these rarely have a solid data-science foundation and often they seem intended to be sensational rather than truly informative.

David's book fills the gap between these categories. It combines practically relevant examples with solid data science fundamentals. My favourite aspect of the book is its collection of real-world vignettes, each illustrating an ethical issue for business (or government). These vignettes reveal a variety of pitfalls that we should be aware of as we incorporate data-science-based techniques in our businesses.

We have heard some of the stories: image classifiers labelling people as gorillas; retailers taking actions because people are predicted to be pregnant, and election campaigns targeting people based on unethically obtained data.

However, how many of us have thought carefully about the actual ethical issues that these cases reveal? (We should.)

David organizes the issues for us and pushes us to think more carefully about them. He takes advantage of the now-well-understood data science process. Doing data science involves: acquiring data, (pre)processing the data, analysing the data, modelling the data, evaluating results, and then using the models/results for insight or to support or make decisions. Separating the ethical issues based on these steps provides order to what otherwise is a daunting array of possible ethical pitfalls.

To be realistic, no one book is going to make you an expert. But we also cannot just hope someone else will take care of it for us. For example, our lawyers might be able to help us with legal and reputational risk, but ethics isn't simply about following the law or managing your reputation. We all need to become students of doing the right thing.

As a student, it helps to have a great teacher. Like David.

Foster Provost
New York, 2021

Preface

Regarding this book

As I started teaching the Data Science and Ethics class in the graduate program of the University of Antwerp, I felt there was a need for a textbook that could guide such courses. I specifically wanted to write a book that is useful to both business and computer science students, as this topic is of importance to both. I believe that his topic is inherently multidisciplinary and hence focus on both the concepts and techniques, as well as the cautionary tales.

Who should read this book?

This book is intended for everyone who wishes to learn about the ethical aspects of data science, including:

- Business students and business people who work with data scientists, or manage data-driven businesses. Managers in many sectors and of various levels need to be able to ask the right questions when it comes to data science, be able to interpret and challenge the results and, based on these, make the right decisions. As the ethical aspects become ever more important, business people need to be aware of the concepts, techniques and cautionary tales of data science ethics. This is what the book aims for: provide guidance and insight on deciding what is right and wrong when conducting data science.
- Data science students and data scientists. A data scientist is more than a human making calls to predefined libraries on existing datasets. At every step in a data science project, from data gathering to model deployment, important decisions are to be made. This book aims to help (aspiring) data scientists to understand how technical choices can have ethical implications (for better and worse), while at the same time understanding the wide societal impact of their work.
- People with a general affinity with technology. The topics that are written about are often discussed in the popular media as well. This book

structures and summarizes the key concepts and cautionary tales of one of the most impact technological innovations of our time, data science.

The book requires some basic knowledge of data science, and hence is well suited to be used within Masters in Data Science, Business Analytics, or any program where introductory data science is part of the curriculum.

More resources on the book can be found online, at www.dsethics.com.

Acknowledgements

Writing this book was quite a journey, in which I was accompanied by many persons. Thanks to everyone who provided feedback, inspiration and encouragement throughout the writing of this book. At the risk of forgetting someone, let me specifically thank Dieter Brughmans, Toon Calders, Theodoros Evgeniou, Sofie Goethals, Travis Greene, Vinayak Javaly, Raphael Mazzine Barbosa De Oliveira, Pieter Leyman, Constant Martens, Bjorge Meulemeester, Stiene Praet, Yanou Ramon and Galit Shmueli. An additional thanks to Sam Pinxteren for the idea for the figure of hashing personal IDs in Chapter 2, and Raphael Mazzine Barbosa De Oliveira for his help with the examples of the XAI methods in Chapter 4. A special thanks to Foster Provost for writing the foreword, and the valuable feedback on earlier drafts. I'm also grateful for the help in the publishing process by Katherine Ward and Charles Bath at Oxford University Press. This book corresponds to what I have been teaching in my Data Science and Ethics course at the University of Antwerp. A special thanks to the students who contributed to the many discussions in that class, which often found their way into this book.

Most importantly, I am very grateful for the support and love of my wife and two kids. I started writing this book when my first son was still in diapers and my second son wasn't even born. You've been an inspiration to me throughout my writing. This book is dedicated to you.

- David Martens

Endorsements

"An excellent reading with both depth and breadth on some of the most important challenges and risks data scientists, businesses, governments and societies face today as Artificial Intelligence adoption grows. These are topics everyone needs to be aware of, and this is one of the very few must read books on these issues"

- Theodoros Evgeniou, Professor of Decision Sciences and Technology Management at INSEAD, France

"This is an important and timely book for data scientists, written in a clear and engaging way. Motivated by many relevant examples, the author successfully de-mystifies data ethics lingo and presents a comprehensive view of ethical considerations during the entire data science lifecycle."

- Galit Shmueli, Tsing Hua Distinguished Professor, Institute of Service Science and Institute Director, College of Technology Management, National Tsing Hua University, Taiwan

Contents

About the Author

The photographer Marc Wallican

David Martens is a Professor of Data Science at the Department of Engineering Management, University of Antwerp, Belgium. He teaches data mining and data science ethics to graduate students studying business economics and business engineering. He has a long track record in explainable AI research, and has won several best paper awards. In his work, David has collaborated with large banks, insurance and telco companies, as well as with various technology startups.

1

Introduction to Data Science Ethics

1.1 The Rise of Data Science (Ethics)

"It is the best of times, it is the worst of times, It was the age of wisdom, it was the age of foolishness."

Charles Dickens, 1859 [114]

In 2010, Mark Zuckerberg was chosen as Person of the Year by *Time* Magazine [182]. The 26-year-old founder of Facebook was celebrated for creating a new way of exchanging information. At that time, Zuckerberg saw the future of Facebook as follows: 'the last five years was the ramping up, I think that the next five years are going to be characterized by widespread acknowledgment by other industries that this is the way that stuff should be and will be better', [182]. Fast forward to April 2018. Zuckerberg is at a congressional hearing, answering questions on the Cambridge Analytica debacle, where the data of approximately 87 million Facebook users were obtained by the outside company, without the explicit permission of the users [426, 199]. The US Federal Trade Commission reportedly imposed a US\$ 5 billion fine to settle the privacy concerns [141, 236]. The company additionally faced push-back on ethical issues as discrimination against sensitive groups in their ad targeting [290, 244], and the involvement in a mood manipulation study [83, 208]. By 2021, Facebook has launched several initiatives to address such ethical issues, including the installment of an independent Oversight Board that issues policy advisory opinions on Facebook's content policies [45, 312, 219], numerous tools such as Fairness Flow [138], 'Privacy Checkup', and 'Why Am I Seeing this Post' [137], and 7.5 million US\$ in funding to create an independent AI ethics research centre in Munich [79], thereby demonstrating the importance of the ethical aspects of data science in the 21st century.

Data science has so far mainly been used for positive outcomes for businesses and society, for example in risk management, to predict terrorist attacks or

Data Science Ethics. David Martens, Oxford University Press.
© David Martens (2022). DOI: 10.1093/oso/9780192847263.003.0001

detect tax fraud, or in a business setting to increase profitability and revenues or reduce costs. Citizens have enjoyed better more efficient services thanks to data science. However, just as with any technology, data science has also come with some negative consequences: an increase of privacy invasion, data-driven discrimination against sensitive groups, and data-driven decision making without explanations.

Ethics is all about what is right and what is wrong. This book looks at the different concepts related to data science ethics, data science techniques that can help with or lead to ethical concerns, and cautionary tales that illustrate the importance and potential impact of data science ethics.

1.2 Why Care?

Data scientist was famously once described as the sexiest job of the 21th century [103]. And indeed, it is a great job to have. You get to be a detective, looking for interesting patterns that can solve real problems. You get to be a designer, looking for creative data sources, features, or use cases. You get to be a rock star, having the ears of a large audience who are in awe of your magical work. You get to be a modern prophet, predicting future outcomes. But maybe most importantly, you get to be impactful, as your work is likely to have an impact on the business's bottom line, and lead to decisions being made for many persons. With these great opportunities also come great responsibilities. A data scientist is more than a human making calls to predefined libraries on existing datasets. At every step in a data science project, from data gathering to model deployment, important decisions are to be made.

Data science ethics is arguably even more important for managers in businesses where data science practices are a key asset. A 2011 McKinsey report foresaw a need of 140,000 to 190,000 more deep analytical talent positions, and 1.5 million more data-savvy managers (on top of the 2.5 million already in place) in the United States alone by 2018 [279]. Although the numbers are outdated, the 1 to 10 ratio is interesting. It does not imply that every data scientist needs 10 managers; rather, it tells us that managers in many sectors and of various levels need to be able to ask the right questions when it comes to data science, be able to interpret and challenge the results, and, based on these, make the right decisions. As the ethical aspects become ever more important, business people need to be aware of the concepts, techniques, and cautionary tales of data science ethics. This is what the book aims for: to provide guidance and insight on deciding what is right and wrong when conducting data science.

Data scientists and managers are not inherently unethical, but at the same time not trained to think this through either. The many cautionary tales will demonstrate how quickly ethical aspects can be overseen. The racist chatbot of Microsoft [337, 359], the wrong prediction of a picture with black people as gorillas by Google Photos [402, 41], the inability to swiftly counter accusations of discrimination against women by Apple Card [5, 318], the apparent discrimination against women by Amazon's predictive recruiting system [102], and the Cambridge Analytica debacle related to Facebook data [112, 380] are just a few illustrations of how even these giant tech companies, with massive data science capabilities, brilliant data scientists, and business people, can be confronted with ethical issues.

You might wonder, why is this important, and why should I care about data science ethics? Although being ethical has been put forward as a life goal in itself [17], there are just as important societal and business reasons. First of all, there are huge reputational and financial risks related to data science ethics. The numerous cases in this book will demonstrate this point. Not only large companies risk their reputation, also startups and smaller companies should care: they often rely even more on new data science products and services. Not getting the ethical aspects right can stop their growth (or even business) altogether or could lead them into trouble during due diligence or investment negotiations. Reputational risks easily translate into financial risks. As unethical data science can lead to mental and physical harm, lawsuits and settlements can result in large financial losses as well.

A second reason to care about data science ethics is the actual value it can bring. Ethical thinking can lead to improvements in your data and data science models, with potentially more accurate predictions or better user acceptance of the data science models. For example, in Chapter 4 we will see how explaining complex prediction models can provide insight on how the model is making mistakes, and how to fix these. Beyond the improved data science models, ethical practices can be a great marketing instrument, similar to how Apple increasingly puts an emphasis on the privacy aspect of its products [342]. Data science ethics can thereby improve the business value, through more revenue, lower costs, or higher profits.

Thirdly, we've reached an age where society expects business leaders and data scientists alike to be ethically responsible. The power of data science has become clear to both the data subjects, data scientists and business leaders. The cases that regularly appear in the media, from privacy-related discussions to real data science cases showing to have unfairly treated certain sensitive groups, have educated the public. Members of generation Z, born between

1995 and 2010, care a lot about social justice and ethics [149], and are putting more emphasis on corporate social responsibility. In any company where data science takes up an important role, data science ethics should have the same attention as other components of Corporate Social Responsibility.

1.3 Right and Wrong

Ethics is a discipline likely as old as philosophy, and is defined as follows:

> **ethics**: 'Moral principles that govern a person's behaviour or the conducting of an activity'.
> **moral**: 'Concerned with the principles of right and wrong behaviour'.
> Oxford English Dictionary[1]

Discussing what entails ethical behaviour can, and does, fill many books. One of the most important philosophical works in ethics is Aristotle's 'Nicomachean Ethic' [412, 17]. Aristotle states that we study ethics in order to improve our lives. Through proper upbringing and teaching, we can find the righteous actions to take, which can lead to right habits. These can in turn lead to a good stable character (which is conscious, unlike habits). In that spirit, this book aims to teach you what ethical data science entails, so you can take righteous data science actions when they are called for.

Another important insight from Aristotle is that moral behaviour can be found at the mean between two extremes, the one is excess, the other being deficiency [412]. Find a moderate position between these two extremes, and you will be acting morally. This 'golden mean' condition is an important concept, which we will also point to when it comes to ethical data science: often the right thing to do balances between using no data at all (deficiency) and the use of all available data for any possible application, without any concern for issues as privacy, discrimination, or transparency (excess).

An interesting distinction can be made with the law: whereas the law tells us what we *can* do, ethics tells us what we *should* do [368]. Ethics answers the question: *What is right and what is wrong?* Although this book is not about the legal aspects of data science, law and ethics are intertwined as ethics sometimes evolve into laws, and as lawful and ethical thinking can overlap. In Europe for example, the General Data Protection Regulation (GDPR) covers many of the

---

[1] https://en.oxforddictionaries.com/definition/ethics
https://en.oxforddictionaries.com/definition/moral

privacy-related and even explainability aspects of data science (as discussed in Chapters 2 and 4, respectively), while at the time of writing this book, the European Commission is even proposing a new regulation on the topic of trust in Artificial Intelligence (AI) [131]. Technological advances in the data science domain are moving so fast that legislation is in a continuous struggle to catch up.

If there are no clear laws guiding us, then who decides what is moral, what is right, and what is wrong? Each person and business will do so on its own, deciding where it would like to be on the continuum between deficiency and excess. This will be influenced by how society and customers value data science ethics. So in fact, you, as a customer and member of society, are deciding on what is right and what is wrong as well. This further reveals the subjectiveness of ethics. Let's illustrate this with an important ethical aspect: discrimination. Data science in itself is all about discriminating: discriminate between the loan applicants that are likely to repay their loan from the ones likely not to, discriminate the likely churners from the likely not churners, discriminate between the prospects likely interested in my product, from the ones likely not interested in my product. But one of the ethical considerations to make is not discriminating against sensitive groups. But who decides what sensitive groups are? Typically, discriminating against race, gender, or religion is considered unfair, making for three important variables to test when considering fairness. But this is not always the case and depends on the application: in medical diagnosis for example, race and gender can be important scientifically motivated variables.

Next to the application-dependency, what is considered sensitive also varies over time and regions. Not discriminating against women for example is a relatively recently accepted standard. In the United States (US), the legal right for women to vote was established nationally in 1920. Only in 1976 did West Point, a US military academy, admit its first female cadets [37]. In Europe, women were allowed to vote for the first time throughout the 20th century [465]. In Belgium for example, a woman's right to vote was established in 1948, while Moldova only provided this right in 1978. Similarly, not discriminating against race is also something that was not always considered the right thing to do. Slavery in the US, primarily of black people, was only prohibited nation-wide in 1865 by the 13th Amendment. While the right for black people to vote was only included in the 15th Amendment in 1870 [215].

Although now, most of us consider these to be self-evident, we too will be considered victims of our time in the future. Two groups emerge. First, those for whom we consider it is not wrong to discriminate against, but who currently do have all the rights that most humans have. Think of elderly people, or people with a lower income. Age surely plays an important role in marketing and insurance decisions. Income certainly does as well, targeting ads towards iPad users being a simple example. Perhaps in the future we will consider this to be unacceptable, making age and income also sensitive variables. Secondly, some groups which we consider now not to be worthy of all the rights that we have might become sensitive groups as well. Think of animals or robots. Figure 1.1 shows how the topic of veganism, where one abstains from using animal products, has increased in popularity over the years. Given this trend, it is plausible that our great-grand-children will condemn the meat-eaters among us as being immoral. Therefore, remind yourself to be mild when you look at the historical cautionary tales, including the ones described in this book, as these tend to be interpreted with newly obtained ethical insights from our current time.

Also regional effects play a role. The difference in US versus European law already demonstrates this. A 2018 study at MIT studied the human preferences when dealing with the trolley problem [28]: if a driver could not break and

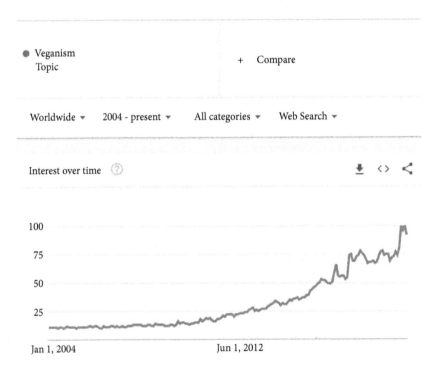

Fig. 1.1 Google Trends for the topic Veganism (up to early 2020), arguing for an increased interest in society for animal rights.

had to choose between hitting and killing an adult versus a baby, what would be preferred? By putting such dilemmas forward to thousands of persons, considering different types of subjects (babies, children, cats, dogs, elderly, executives, homeless, etc.) a ranking of ethical preferences could be established. The study found that, in terms of sparing, children were preferred over adults, adults over elderly people, and interestingly dogs over criminals. Important regional differences were also found in this study: the preference to spare young over older characters was much less pronounced in Eastern countries, such as Japan and Taiwan, as compared to Western countries, such as North America and Europe. Similarly did the authors find that Latin-American countries have a weaker preference to spare humans over pets. This highlights regional differences related to respect for the elderly or animals. Respect for the rights of individuals versus the state also differs across countries, as demonstrated for example by the different privacy regulations in Europe, the US and China [346].

Given the subjectiveness of ethics, deciding on what data science ethics practices to implement in your business is an endeavour that every organization has to undertake. The book can help both managers and data scientists in this decision process by pointing to the concepts and techniques to consider, and the cautionary tales to remember.

1.4 Data Science

There is quite some confusion in the domain with regards to terminology. Therefore several definitions are provided, as to ensure that the reader understands the context and key concepts.

> **data**: 'Facts or information, especially when examined and used to find out things or to make decisions'.
> **algorithm**: 'A set of rules that must be followed when solving a particular problem'.
> Oxford English Dictionary[2]
> **prediction or AI model**: 'The decision-making formula, which has been learnt from data by a prediction/AI algorithm'.

---

[2] https://en.oxforddictionaries.com/definition/data
https://en.oxforddictionaries.com/definition/algorithm

The key building block of all data science is data. On such data, algorithms can be applied that lead to certain data science models. One can argue which of these can be unethical, data surely can be: if it includes personal data that a data subject doesn't want you to have, or if there is a bias against certain sensitive groups. A predictive model surely can also be unethical, most often because it was built on data which has ethical issues. A predictive model that makes use of personal information, or makes predictions that discriminate negatively against sensitive groups for example. An algorithm is nothing more than a set of rules or steps that are to be followed, labelling it as ethical or unethical is less straightforward (unless the developer of the algorithm has explicitly included ethical or unethical aspects). Consider a decision tree algorithm that is applied to data with privacy issues, which will yield a prediction model with privacy issues, in line with the often-used data quality statement: Garbage In, Garbage Out. The data and the resulting prediction model are unethical, but therefore calling a decision tree algorithm unethical does not seem warranted. The first of Kranzberg's six laws of technology similarly states: 'Technology is neither good nor bad; nor is it neutral.' [248]

When we speak of data, several types of data merit special attention: personal data, which is important when considering privacy, and sensitive data, which requires additional care for privacy but which also should not be used to discriminate (in most cases, remember the medical diagnosis domain where such data could be useful). Behavioural data finally is an increasingly available data source, which provides digital breadcrumbs as we move through the world.

Personal data: "'personal data' means any information relating to an identified or identifiable natural person ('data subject'); an identifiable natural person is one who can be identified, directly or indirectly, in particular by reference to an identifier such as a name, an identification number, location data, an online identifier or to one or more factors specific to the physical, physiological, genetic, mental, economic, cultural or social identity of that natural person'.
(GDPR, Article 4)
Sensitive data: 'personal data revealing racial or ethnic origin, political opinions, religious or philosophical beliefs, or trade union membership, and the processing of genetic data, biometric data for the puse of uniquely identifying a natural person, data concerning health or data concerning a

natural person's sex life or sexual orientation shall be prohibited'.
(GDPR, Article 9)
General Data Protection Regulation
Behavioural data: 'Evidence of actions taken by persons'.
Shmueli (2017) [396]

The terms used for the domain of algorithms and applications when working on data have evolved over time, from business intelligence to analytics, data science, and AI.

data science: 'A set of fundamental principles that support and guide the principled extraction of information and knowledge from data'.
data mining: 'The actual extraction of knowledge from data via technologies that incorporate these [data science] principles'.
artificial intelligence: 'Methods for improving the knowledge or performance of an intelligent agent over time, in response to the agent's experience in the world'.
Provost and Fawcett (2013) [363]

Many of the discussions within data science ethics are on predictive modelling or supervised learning, in which data mining (or machine learning) techniques are used to find patterns in the data, in the form of a prediction model that predicts the value of some target variable. Descriptive modelling or unsupervised learning will also extract patterns from data, but these patterns are not (explicitly) used to make predictions, but rather to discover descriptive patterns in the data. Clustering and association rule mining are popular descriptive data mining tasks. Though the focus is often on supervised learning, unsupervised learning might just as well have the same ethical issues.

Artificial Intelligence is historically defined as the domain of intelligent agents, where computers attempt to mimic human cognitive functions such as learning and problem solving. The domain of data mining is obviously a part of this: having a computer learn patterns from data, that can be used to solve certain problems. The recent rise in popularity of (the term) AI is mostly due to deep learning [262]. The success of these large artificial neural networks can be attributed to the increased availability of data, processing power, and methodological improvements. Mainly in image and speech recognition does deep learning yield tremendous results, with sometimes super-human results that are being reported. A major ethical issue with such models is that they are typically black box and provide no explanation why a certain prediction is made. The term of AI is often used for both data science and data mining. As data science is the overarching term, we will mostly use this term in the rest of the book. Just be aware that the term AI might also be used (and who knows what the next buzzword will be) when discussing the same cases. A part of AI lies beyond data science, such as Artificial General Intelligence and singularity. The specific ethical aspects related to this are likely not to be so relevant in our everyday lives in the near future.

1.5 Data Science Ethics Equilibrium

A recurring theme in the book is that ethical thinking is not boolean: a data science practice is not simply ethical or unethical, most often it is a continuum that balances ethical concerns and utility of data (science). At one extreme, there is no investment at all, nor interest, in data science ethics, while at the other extreme the ethical concerns are so overwhelming that no data is being used. Most data science applications will care about both to some extent. Already notice the similarity with Aristotle's belief that ethical virtues are often found by some average in between two extremes. So the extreme of not doing anything with data, under any circumstances, can also be deemed unethical.

This balancing act between ethical concerns and utility is illustrated in Figure 1.2. Depending on the importance of ethical concerns and utility of the data, data science practices are determined. The obtained equilibrium is very much context-dependent, and is determined by the potential impact on humans and society, and the extent to which this impact is deemed right or

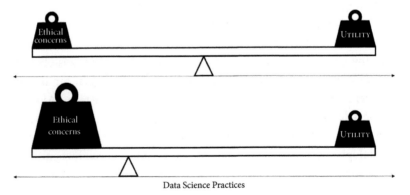

Data Science Practices

Fig. 1.2 Data Science Ethics Equilibrium.

wrong. Some example impact questions include: How much valuable data is available? How important is data science for my company? Am I using personal data? Do the data-driven decisions have an impact on persons, and if so, on how many? The importance of ethics to your setting is subjective, as discussed previously, and begs such questions as: Do my customers care about the ethical issues of my data science practices? Do my shareholders or other stakeholders care? Does the CEO or data scientist understand and care about data science ethics? Etc. The 2021 newly proposed rules by Europe for Artificial Intelligence follow a similar risk-based approach [131]. For example, AI systems that threaten the safety of people are considered to have an unacceptable risk and will simply be banned. High-risk systems include those that are used in employment, law enforcement or migration, and come with strict obligations. The equilibrium will lie closer to the left and hence require more stringent data science ethics practices to be included. The final two categories that are defined in the proposed European rules are limited risk (e.g. chatbots) and minimal risk (e.g. spam filters).

We shouldn't be naive: some businesses will simply don't care much, and might argue 'They won't find out anyway.' In that sense, the importance of data science ethics in a company is related to the values of a company. If values such as ethical thinking, transparency, a customer-first mindset, or leading the industry are not considered important, data science ethics is likely to take up a small role as well. The ethical equilibrium then determines the data science practices: the more to the left side of the spectrum, the stronger the need for the data science ethics practices that will be discussed in this book, such as limiting the data that can be used to non-personal data, actually removing signal from the data by adding noise or generalizing variables, changing the labels so as to remove potential discrimination against sensitive groups, requiring more evaluation analysis to be done, restricting the space of predictive models to those that are comprehensible, etc.

> **Data Science Ethics Equilibrium**: 'A state of data science practices determined by the ethical concerns and utility of data science'.

Consider for example credit scoring, where banks use data science to build prediction models that assess the creditworthiness of loan applicants [446]. The decision on granting or denying credit can have a large impact on the loan applicants, and the bank has plenty of personal and sensitive data, from income data to a record of all payments that were done by its customers. In this case, the ethical concerns are important, requiring that the privacy is protected, that the credit granting decisions can be explained to the loan applicants, and that there is no discrimination against sensitive groups. At the same time, the utility of data science is massive, as indicated by the maturity of data science in banking. This leads to an equilibrium of data science somewhat similar to that of the top of Figure 1.2, with practices that typically do not use gender in the prediction model, include inherently comprehensible models and have stringent privacy policies. On the other hand, consider predictive maintenance [192], an industry application where data science is used to predict when a machine will break down, in order to send a technician to maintain the machine before it breaks down. The data it uses are typically machine-related data, such as temperature, vibrations, time running, etc. These are of little ethical importance, while the data can be of great utility for this setting. The equilibrium then quickly shifts to the utility side of the continuum, with much less ethical concern. Now consider that also data is being used on the persons that maintain and come near the machine. Data on these persons that are available include their name, gender, time of working, and nationality. This emphasizes the ethical concerns, thereby shifting the equilibrium again slightly to the left, with changing data science ethics practices.

The danger exists in not thinking this balancing act through, often leading to a non-equilibrium, as shown in Figure 1.3. The ethical constraints are of course not removed, they are simply ignored, very likely leading to pushback later on against the implemented data science practices. Thinking these ethical issues through is not an easy task, and is exactly what this book tries to help you with. Ethics requires to discuss this balancing act. The numerous Discussions in the book will provide structured exercises to do so. Additionally, most of the cautionary tales are examples of such a lack of ethical equilibrium. Be aware though, that these tales mostly are about technological innovations on the forefront of the field at that moment in time. The ethical implications might seem obvious in hindsight, but often are not at the moment of initial implementation.

Whereas other books have demonstrated the utility of data science (see e.g. [364, 398, 29, 423]), this book will guide you in understanding what the ethical concepts are, how important these can be to your setting, and teach you

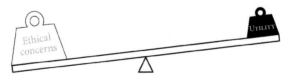

Fig. 1.3 Lack of a Data Science Ethics Equilibrium.

data science ethics techniques that can be used in your data science practices, to balance the utility and ethics of your data science application. Finally, note that addressing all issues for 100% is close to impossible (as will become clear in the next chapters): for example, secure storage and communication are hardly ever fully guaranteed, how to explain black box models and how to evaluate such explanations are not yet perfectly defined, and avoiding discrimination in data science models exists in different, often conflicting, objectives. Unrealistic demands will likely lead to the extreme of not using any data at all under any circumstances, which Aristotle might have argued to be unethical as well.

1.6 The FAT Flow Framework for Data Science Ethics

The increased attention and acknowledgment of the importance of ethical data science is illustrated by the attention from mainstream press on the cautionary tales, and the ample new technical research being proposed, from new approaches on how to explain prediction models, to studies discussing whether to include ethical preferences in self-driving cars. This book will be structured around the 'FAT Flow Framework', using three dimensions:

1. Stage in the data science process
2. Evaluation criterion
3. Role of the human

The first dimension follows the five common (and relevant) phases of a data science project: from data gathering to model deployment. The second dimension entails three properties that have emerged in the community: fair, accountable, and transparent (FAT). The third dimension of the framework considers the four roles for humans that exist when discussing data science ethics: data subject, data scientist, manager, and model subject. This framework is general enough to include most known ethical aspects of data science, and the flexibility to include novel techniques and cases. At the same time it provides a scientific instrument to follow with guidelines and cautionary tales.

The present discussion of the framework is not claimed to be complete, as any framework or checklist is bound to become outdated quickly (and hence needs to be updated regularly): new data sources, techniques, applications, and ethical considerations are continuously being proposed. Rather, data scientists

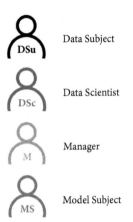

Data Subject

Data Scientist

Manager

Model Subject

Fig. 1.4 The different roles in data science projects.

and businesses can use the FAT Flow Framework as a generic guide at the start of a data science project, and when reviewing existing data science projects.

1.6.1 The Different Roles in Data Science

Humans enter the data science process in different roles, as discussed by Martens and Provost [287]. One typically only considers the data subject role. Even though this indeed is one of the crucial roles, there are other roles as well that can require different considerations:

1. Data Subject: the person whose (personal) data is being used. A regulator can act as a proxy.
2. Manager: the person who manages and/or signs off on a data science project.
3. Data Scientist: the person who is performing the data science.
4. Model Subject: the person on whom the model is being applied.

How this role differs can be illustrated with the issue of being able to explain the decisions made by some data-driven credit scoring model (an issue we'll discuss at length in Section 4.4), see also Table 1.1 Such a model will predict whether a loan applicant is able to repay his or her loan, and hence whether to grant credit or not. First of all, a rejected loan applicant will want to know why the application is rejected (even more: it often is a legal requirement). Is it because the income is too low? Because the ratio of loan to value of the mortgage is too high? Is there some issue in the credit history? Is it a combination of factors? Simply stating 'Computer says no ...' is not enough.

Table 1.1 Description of potential explanations needed in a credit scoring context, depending on the role of the person asking for an explanation.

Role	Credit scoring role	Relevant explanation
Data subject	Customer	How is my data used?
Data scientist	Data scientist	Why this wrong prediction?
Manager	Risk manager	How does the model generally work?
Model subject	Loan applicant	Why was credit denied?

Second, before a credit scoring model is being deployed, the manager of the bank will want some insight into how the model works. Simply deploying some black box, incomprehensible model will not be allowed, even if it is accurate on an out-of-time test set. For the manager it is less important to know why a single customer is accepted or rejected, the manager will want the general idea of how it works. Third, we have the data scientist. He or she will be eager to know why the model is making certain wrong decisions. Is it because not enough data was available on that specific group? Is it because of a data quality issue? Knowing this can help to improve the data science model. Finally, note that in this case the data subjects are different from the loan applicants: the data subjects are all customers who previously took out a loan. On these persons the bank knows the actual target variable: did this customer repay the loan or not. The loan applicant is the one applying for a loan, being scored by a model built on the data of persons who took out a loan previously. So the data subject in this case does not really need an explanation for the prediction, as it has little impact on him or her (except for the fact that his or her data is being used, but that is a different matter).

When we consider the different stages and criteria of the framework, some roles become more prominent than others, and sometimes even ask for different treatment, as demonstrated with the credit scoring explanations.

1.6.2 FAT: Fair, Accountable, and Transparent

Data science ethics can be evaluated using three criteria: fairness, transparency, and accountability. The first two criteria evaluate ethical *concepts*, such as privacy, discrimination, and explainability. Accountability is about the effective and verifiable *implementation* of these concepts. As 'FTA' doesn't sound as nice, these criteria are usually abbreviated as 'FAT'. Next follows a formal definition and explanation of each criterion.

Fairness covers two important concepts: discrimination and privacy.

> **fair (1):** 'Treating people equally without favouritism or discrimination'.
> **fair (2):** 'Acceptable and appropriate in a particular situation'.
> Oxford English Dictionary[3]

As touched upon, both are too general of a definition: the point of many data science models is to discriminate between groups: the good and bad payers, the churning and loyal customers, the likely interested versus not interested persons, etc. Similarly does the 'particular situation' at hand need to be specified. Therefore, the definitions we will use are as follows:

> **fair (1):** 'Not discriminating against sensitive groups'.
> **fair (2):** 'Acceptable treatment of privacy aspects'.

Next to the discrimination aspect of fairness, there is also the privacy aspect. The fair use of personal data entails that the privacy of the data subject is respected.

Privacy is recognized as a human right. The United Nations' 1984 Universal Declaration of Human Rights states: 'No one shall be subjected to arbitrary interference with his privacy, family, home or correspondence, nor to attacks upon his honour and reputation.' The European Convention on Human Rights of 1953 writes on privacy: 'Everyone has the right to respect for his private and family life, his home and his correspondence.'

Much has been written about privacy, from Orwell's book *1984* to the European GDPR regulation [335]. But what is privacy?

> **privacy:** 'A state in which one is not observed or disturbed by other people'.
> Oxford English Dictionary[4]

In other words, fairness also relates to an acceptable treatment of individuals with respect to their privacy, not observing or disturbing people when they don't want to be. This too might be self-evident, but once more this is not a boolean criterion. Some applications allow for more latitude: for detecting fraud, solving crimes or medical diagnosis, generally more personal data can

---

[3] https://en.oxforddictionaries.com/definition/fair
[4] https://en.oxforddictionaries.com/definition/privacy

be used (of course still with some limits) as compared to targeted advertising or music recommendation. Regional expectations on privacy also differ, as illustrated by the relatively strict regulations of the notable GDPR in Europe for example [335, 343].

The transparency criterion is probably the most important one, as it has implications for both accountability and fairness. Although one often limits transparency to explaining the decisions made by the data science model, transparency goes much further, covering all stages of a data science project.

---

transparent: 'Easy to perceive or detect'.
Oxford English Dictionary[5]

The transparency required is also different according to the different entities to which transparency is provided: the manager of the organization will want full transparency on the process, which the data subject might not get (so as not to reveal company secrets). On the other hand, the data scientist will want full transparency on all algorithmic steps taken in previous modelling exercises; the manager is likely not that interested in knowing what hyperparameter grid was used in cross-validating the regularization hyperparameter of a regularized logistic model. In the next sections we will discuss which transparency is required and how to provide this, based on the stage and data science role.

Transparency also takes an important role in accountability and fairness. In order to know whether a model is fair, transparency is needed in the data used (privacy) or the evaluation of the model for different sensitive groups (discrimination). As we will see, making the data science process and predictions transparent, the data science model might even improve: unnecessary data sources might be detected, data biases which led to decreased performance can be removed or explanations of misclassifications can reveal such insights as how to improve the data quality or model.

An important aspect of transparency is the ability to explain the decisions made by the data science model to model subjects. There are legal scholars who argue the existence of a 'right to explanation' in the European GDPR [21]. Article 14.2.g requires that data subjects not only have the right to know that there is automated decision making, including profiling, but also that the data subject then has the right to obtain meaningful information about the logic involved. The advisory organ of the European Union on GDPR, Working Party

---

[5] https://en.oxforddictionaries.com/definition/transparent

29, provides additional details on this concept of 'logic involved'. They write: 'The controller should find simple ways to tell the data subject about the rationale behind, or the criteria relied on in reaching the decision. The GDPR requires the controller to provide meaningful information about the logic involved, …The information provided should be sufficiently comprehensive for the data subject to understand the reasons for the decision.' [22] This does not require businesses to disclose every detail to the model subjects. Doing so could infringe on company secrets or even the privacy of data subjects. What exact data science technique you are using could be part of your 'secret sauce', just as you don't necessarily have to provide the model subject the exact prediction score of your model. This illustrates that transparency as well is not a boolean criterion, but more a continuous one.

The two components of transparency can be defined as followed:

> **transparency (1)**: 'Clarity in the data science process'.
> **transparency (1)**: 'Ability to explain decisions made by data science models'.

Accountability is the third, and might well be the least well-defined FAT evaluation criterion, mainly because it is a very broad concept, often used in different contexts.

> **accountable**: 'Required or expected to justify actions or decisions; responsible'.
> Oxford English Dictionary[5]

In a paper on public accountability, Mark Bovens talks about the 'elusive concept of accountability' [61]. In an opinion paper by the data protection working party on accountability, it is stated that 'even though defining what exactly accountability means in practice is complex' [19], while Charles Raab talks about the concept in the context of privacy protection, as 'The question of accountability … is highly complex with deep implications for the relationship between organisations and the public' [369].

The GDPR text and supporting documents are very useful to help us to understand *why* accountability is needed (see Opinion 3/2010 on the principle of accountability 2010 [19]): accountability is all about moving from theory to practice. Policies alone are not enough. A recitation of a company's policy with

---

[5] https://en.oxforddictionaries.com/definition/accountable

regards to data science ethics can be very lofty and impressive, but there needs to be obligations to ensure that these are put into practice. Accountability is intended to strengthen the responsibility of a company and its people, leading us to another definition:

> **responsible**: 'Having an obligation to do something, or having control over or care for someone, as part of one's job or role'.
> Oxford English Dictionary[7]

This obligation is crucial. Having to account to someone for one's actions (or failure to perform certain actions) makes an important difference [311]. This obligation has three components.

> **accountability**: 'Obligation to (1) implement appropriate and effective measures to ensure that principles are complied with, (2) demonstrate compliance of the measures upon request, and (3) recognize potential negative consequences'.

The first term is about taking appropriate *measures* [19] to ensure that data science is conducted according to the set policies and relevant regulations. Technical measures include the use of proper encryption methods to ensure privacy, or generally the consideration of the discussed data science ethics techniques throughout the process. Organizational measures include aspects such as training on these matters and oversight, which we will discuss in Chapter 6 when discussing the ethical issues relating to model deployment.

The second term of accountability is being able to *demonstrate* that these measures are implemented, and that these do indeed lead to compliance with the set forth policies and regulations [19]. The ability to demonstrate what measures have been taken is important in this account, and involves a description of tools and training that are put in place, as well as a system for oversight and verification. In other words, accountability does not wait for a system failure [369], rather it requires that an organization is prepared to demonstrate upon request that it is complying with data science ethics. Often this verification is done by proxies of the general public, such as supervisory authorities or auditing agents.

---

[7] https://en.oxforddictionaries.com/definition/responsible

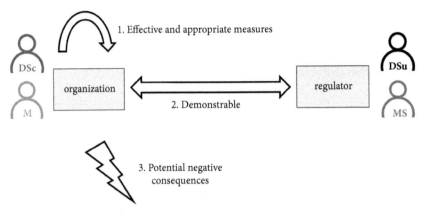

Fig. 1.5 Three facets of accountability.

The third aspect of accountability is having to face possible consequences for not complying with the obligations. When something does go wrong, accountability will be about this aspect mostly, and implies a liability. Financial negative consequences can be in the form of government fines or compensations awarded by judges. The consequences can go much further, as also disciplinary or reputational consequences can be harmful to the organization. As motivated by Bovens [61], this possibility of sanctions (and not necessarily the actual imposition of sanctions) makes the difference between non-committal provision of information and actually being held to account. The three components of accountability are summarized in Figure 1.5.

An interesting link can also be made with financial accounting of organizations to shareholders [369]. These annual reports also tell an 'account' or story about the company. It explains successes and failures, goals and strategy. Depending on the size and type of company, these statements are independently audited, found to comply with certain kinds of accounting standards and conventions. Shareholders for publicly traded companies can also ask questions to the company, and demand remedies. The potential for negative consequences in case of wrongdoing is, of course, very much present as well.

1.6.3 FAT Flow Framework for Data Science Ethics

The FAT Flow framework provides a guide for data science projects. The three dimensions provide the granularity to cover all current and future aspects of data science ethics. Figure 1.6 illustrates the framework, according to (1) the

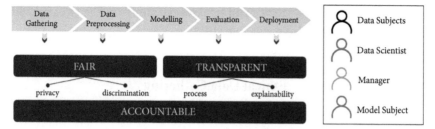

Fig. 1.6 FAT Flow Framework for Data Science Ethics, using three dimensions: (1) modelling stage, (2) evaluation criterion, and (3) role.

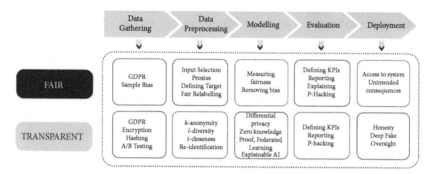

Fig. 1.7 Concepts of Fairness and Transparency within the FAT Flow Framework.

stage of the data science project[8], (2) the FAT evaluation criterion to consider, and (3) the involvement of different roles. Figure 1.7 summarizes the fairness and transparency concepts that will be covered in this book, in the different stages of a data science project. The accountability aspects are partly considered in the deployment stage, and require the actual implementation of effective and demonstrable measures that include the concepts and techniques related to fairness and transparency.

It is important to emphasize that the framework does not provide 'crisp' answers on what ethical concepts and techniques to apply. For all three FAT criteria, there is a range of solutions. Some legal requirements might exist, which make for necessary conditions. The extent to which one addresses the different issues should also depend on the business and data situation: a small startup will have less extensive measures in place than a large multinational

---

[8] CRISP-DM is another popular framework that describes the phases of a data science project [393]. It differs from the one proposed in several ways, for example CRISP-DM does not consider a data gathering phase, while FAT Flow considers business understanding a prerequisite and part of all other phases.

who has dedicated data science and legal teams. Similarly, the size, risk, and sensitivity of the data also play a role [19].

Ethical Data Gathering

Fair
The data gathering process is a first important stage. This needs to be fair to the data subjects and model subjects, in terms of privacy and discrimination against sensitive groups. Note that the roles of data subject and model subject can be different in this case, as different data can be used on each in the data gathering process. Let's revisit the credit scoring case, where data of the customers who ever have been granted a loan is kept, so as to build credit scoring models with target variable: did this customer repay the loan or not? These persons become customers at the start of the loan (if they weren't customers before) and sign a contract which discusses what data is used, for what purposes, how long the data is stored, etc. Loan applications on the other hand will include many persons who are not yet customers at the bank. They just provide their details (such as income, age, profession, etc.) to be scored and obtain an approval and/or interest rate. But is this personal data only used to score the applicant, or is this data kept for other data science modelling? If the latter, it is important to also obtain an informed consent. The fairness should not only cover privacy; also fairness towards sensitive groups should already be considered. This is to avoid a worse prediction performance on that group later on.

The following questions need to be considered:

- **Fair** to the **data subject** and **model subject**: is the privacy of the data subject and model subject respected. when gathering their data?
- **Fair** to the **model subject**: is a sufficient sample included for all sensitive groups?

The accountability criterion will require appropriate and effective measures to comply with the answers to the questions in the previous paragraph, such as data minimization, complaint and rectification handling procedures, and informed consent.

Transparent
The transparency of the data gathering process also needs to consider the privacy of the data subjects and the model subjects. This includes informed

consent: is the data subject and model subject informed about the data gathering and is consent provided? There should also be transparency in what data is gathered, for what purpose and for how long. A specific setting in which data is gathered across all model subjects is A/B testing. The fact that one is part of an experiment can be sensitive and requires transparency. Suppose your water company is testing a substance that makes people miserable (the business case could be far-fetched). They are proceeding with an A/B test, where half of its customers are receiving water without the substance, and half water with the miserable-making substance. A year later, you realize you were in the second group. How would you feel? Do you think the water company should have informed you? You might think this is an implausible scenario, but what if the experiment was online and more subtle, for example simply by filtering the information to show you? A/B testing is widespread and it is important to make model subjects aware of the fact that they are part of this A/B test, definitely if the test can have an important impact on the test subjects. This of course does not mean you need to tell them which group they are part of.

Also the data scientist and manager require transparency, so as to understand how the data is gathered. The data scientist needs to understand the data gathering process to ensure data quality and perform suitable data preprocessing and modelling, while the manager is the one signing off on the process, so he or she surely wants to know how this all occurs.

- **Transparent** to the **data subject** and **model subject**: what data is used, for what purposes, and for how long?
- **Transparent** to the **model subject**: if A/B testing is performed, is the user aware of this and provided have they informed consent?
- **Transparent** to the **data scientist**: how is the data gathered? Was specific over- or undersampling of certain groups considered?
- **Transparent** to the **manager**: how is the data gathered?

Ethical Data Preprocessing

Fair

People with big shoes tend to live shorter. Let that fact sink in for a moment, and consider why this would be so. Perhaps because they need to spend more on shoes? Because they are more susceptible to falling? No, simply because men tend to have larger shoe size and women live longer than men. This example of correlation not implying causation also shows the issue of correlated variables.

Not including a sensitive variable such as race does not necessarily imply that you are not discriminating against racially sensitive groups. Any variable that is strongly correlated with race might lead to the same discrimination. So if a bank is not allowed to use gender in its credit scoring process, is it allowed to use shoe size? A banker will likely not ask you for your shoe size (beware of bankers that do) but he or she will ask you for your address, and this can be a proxy for race. As we will discuss in Section 4.3, racially defined neighbourhoods were used in the past to deny services, in a practice called redlining.

Next to this fairness issue at the feature dimension (on discrimination), fairness also relates to the instance dimension (on privacy). When you claim to have anonymized your data, are you sure that the data cannot be de-identified? Anonymizing your data can be very useful, as you can continue to work on it for aggregated analytics or to to work on fine-grained data that doesn't include any personal information. The latter use case is also very convenient to foster academic research by making available datasets, as is often done through systems such as Kaggle. However, as we'll see, there is a distinction to be made between pseudononymizing, where personal identifiers are removed, and anonymizing, where no information in the dataset can be brought back to an individual.

In 2006, AOL published a dataset with search queries [186, 478]. The dataset was thought to be 'anonymized', by removing IP addresses and user names, and gave each user a randomly generated user ID, together with a list of the search queries for that user. However, some users were able to be re-identified (including a 62-year-old widow), and some search queries were found to be quite worrisome ('how to kill you wife' [including the typo]). This case, as well as a similar Netflix dataset and thought experiments are elaborated on in Section 3.2.

A third ethical issue in data preprocessing is the definition of the target variable. How to define when a loan is in default is not that straightforward: does this occur when someone misses one payment? Two? What if the person happened to have been on vacation and forgot to make the transfer, or forgot to ensure that sufficient funds were available on the account? The Basel II Accords tell us that we should consider three consecutive non-payments as a default [446]. Now consider an HR analytics application, where we want to predict which job applicants to hire. Using a historical dataset, data mining models can be built that predict a 'successful' hire. But how to define this? Is this anyone who stayed with the company for at least five years? Employees that were promoted to a managerial level? Is it just anyone who got hired in the past? All of these can have ethical repercussions: if the company had a bias against women for such positions, the model will include this bias as well. If female employees took maternity leave during their first years at the company, is this accounted for in the definition of 'successful' hire?

- **Fair** to the **data subject** and **model subject**: if the data is kept anonymized, can the data not be de-identified? Is any sensitive information potentially revealed?
- **Fair** to the **data subject** and **model subject**: are the included variables not indications of sensitive groups, and not correlated proxies that discriminate against sensitive groups?
- **Fair** to the **data subject** and **model subject**: is the target variable defined in a manner that does not discriminate against sensitive groups?

Transparent

The transparancy aspects of the data preprocessing phase concern the open communication of the aforementioned issues on instance, input variable and target variable level. If the data is pseudonymized instead of anonymized, do the data subjects and model subjects know their data is still being kept? Is there sufficient effort being made to ensure the data is truly anonymous? The input selection procedure should similarly be well-documented, and be clear to (future) data scientists and managers, as to ensure that the potential ethical motivations for removing (or keeping) variables are well understood. Finally, the definition of the target variable should be made transparent to all, ensuring that everyone agrees on the definition and practical scenarios that could have impacted the measurement thereof.

- **Transparent** to the **data subject** and **model subject**: if the data is only pseudononymized, are the data subjects made aware?
- **Transparent** to the **data scientist**: what procedure was used to anonymize the data, and how is anonymity measured?
- **Transparent** to the **data scientist** and **manager**: what type of data is being used as input variables? Do any of the input variables relate to sensitive variables?
- **Transparent** to the **data scientist** and **manager**: how is the target variable defined? What are the practical implications of this measurement?

Ethical Modelling

Fair

There are ways to include privacy within data science modelling. Consider the case where you are the data scientist asked to build a variety of predictive models on datasets gathered from several data providers. First of all, personal data on the data subjects is not required for you as these won't be a part of your input set. The name of a person, exact date of birth, or social security number are not (or should not be) relevant in your exercise.[9] In other words, personally identifiable information should be protected and made invisible. Secondly, we need to avoid that sensitive variables can be predicted from the datasets. Perhaps political preference is not explicitly included in the dataset, but could be easily predicted. Examples using Facebook data will illustrate this point [246]. In this case, we want to protect against patterns that should not be learned. At the same time of course, you want to have well-performing prediction models. So the important patterns, of relevance for your business case, should not be removed or degraded in the process. This becomes an even bigger practical problem when databases from several sources are being shared, or when data is being published to a wide public.

In the modelling phase, specific ethical preferences could be included in the model. The main reasons to do so are because the data do not sufficiently reflect the desired outcomes. This could be helpful in cases of wanted positive discrimination towards certain groups, or when rare situations need to be addressed. A data science model that predicts how to drive might at some point need to decide whether to run over an elderly man or a pregnant woman. What the preference is should clearly be an ethical discussion. Yet, these are the low-frequency events that will not or rarely be reflected in the data. We'll discuss whether and how to include such ethical preferences in Chapter 4, using the MIT Moral Machine study, and making the link with the widely studied trolley problem.

- **Fair** to the **data subject** and **model subject**: is irrelevant personally identifiable information being protected?
- **Fair** to the **data subject** and **model subject**: can sensitive attributes not be learned from the data?
- **Fair** to the **data subject** and **model subject**: can ethical preferences be included in the model?

[9] In some specific cases this might be relevant, for example in fraud detection to find patterns in the stated date of birth.

Transparent

The arsenal of modelling techniques available to a data scientist is very broad, from the 'workhorse' logistic regression to advanced deep learning algorithms. One dimension on which these algorithms differ is how comprehensible the generated models are. We typically speak of white box versus black box models [286]. As so often, comprehensibility is a continuous measurement, where rule or tree-based models are often regarded as very comprehensible, while non-linear techniques are considered much less comprehensible. The choice of number of inputs (or regularization) and type of algorithm therefore have an important impact on the transparency of the model. The actual generation of explanations for a decision is closely related to this, commonly known as part of explainable AI.

- **Transparent** to the **data scientist** and **manager**: does the data science model need to be comprehensible or explained?

1.6.4 Ethical Model Evaluation

Fair

In the evaluation phase, the data science model is being evaluated on the aforementioned fairness criteria related to privacy and discrimination against sensitive groups. In Chapter 5, we'll review a range of techniques that measure the extent to which privacy is respected by the model.

When we want to determine whether the model discriminates or not against sensitive groups, a paradox is revealed: the need for the sensitive attribute. To evaluate whether a model does not discriminate on race, the race variable would need to be available to assess its impact on the predictions. If this is available, several techniques have been put forward to evaluate the fairness. But when we don't have the sensitive variable, the situation becomes much more difficult.

- **Fair** to the **model subject**: how much private information is used or predicted?
- **Fair** to the **model subject**: is the model not discriminating against sensitive groups?

Transparent

When evaluating the model, transparency has several important roles. The first is on doing proper model evaluation. It is so important that it even has an ethical implication. Different performance metrics can lead to the different decisions on the model. Consider the example of predicting stock prices. A modeller stating 90% accuracy should lead to questioning remarks. One easy explanation could be that the evaluation was done in 10 days only, during which the stock market increased every day. Simply always predicting an increase leads to a 100% accuracy. A well-performing model though? No. A good data scientist would not make such a mistake. But the intuitiveness of accuracy can sometimes lead to reporting this 'easy' accuracy metric, even when it is not suitable.

- **Transparent** to the **manager**: are appropriate predictive performance measures reported?
- **Transparent** to the **manager, data scientist**, and **model subject**: are the results interpreted correctly?

Ethical Model Deployment

Fair
In deployment of data science systems, decisions are sometimes made on who gets access to the data science model. Who are you considering to score? In credit scoring, banks are often limited to those who have a previous credit history or at least a checking account. It's important to be aware of who exactly has access to the system and who doesn't. Censorship is a specific, deliberate case of providing limited access to your data science system. The case of predicting pregnancy by a large retail chain will provide an illustration on when not to provide access to the system to everyone, in Chapter 6. Finally, the filter bubble that is created by only showing model subjects information and news that is predicted to be of interest to them is a demonstration on when to consider not to only use the data science system for the decisions.

- **Fair** to the **model subject**: who are you and are you not giving access to the data science system?
- **Fair** to the **model subject**: is overruling possible, and to what extent?

Transparent
The way a system behaves in production versus during development can be quite different. Unintended consequences are, by definition, unintended, yet can already be thought off during the design. The Tay AI chat bot of Microsoft that tweeted out some racist comments [337, 359] shows the big implications of

not foreseeing such issues at deployment. One should be transparent on what the unintended consequences might be and how these could be mitigated.

Misleading people through data science models is an ethically very questionable practice. DeepFake is such a technology, based on deep learning, to create real-looking yet fake videos. Transparency in disclosing clearly when fake footage is being distributed is required.

- **Transparent** to the **model subject**: did you consider potential negative unintended consequences?
- **Transparent** to the **model subject**: will people not be misled by your data science model?

1.7 Summary

Data science ethics is all about what is right and wrong when conducting data science. What is ethical changes over time, application, and even region. Every business and person needs to determine the importance of ethical concerns versus the utility of the data for their specific application. These respective weights then lead to a data science ethics equilibrium that determines the right data science practices. This chapter gave an introduction of the key concepts of data science ethics and the FAT Flow Framework, which will structure the content of this book. Next, the very first stage is discussed: gathering data.

2

Ethical Data Gathering

Jenny is the founder of a startup that developed an app for music events.[1]
The app has a login module, where users can login through their Facebook
profile. The user's name, email address, and Facebook likes are used by the
app for a better user experience of the app, and to recommend other music
events. Jenny realizes that the data that she gathered on the users of the app
can be of great value elsewhere, and asks her product manager to look for additional business opportunities. He comes back with several scenarios. First,
the Facebook like data would be useful to predict product interest and hence
for targeted online advertising. After discussing this with an ad tech company, they propose to buy the data. Second, music producers are interested
in buying the data as well, as they would like to know how often and how long
people listen to the music under their licence. To make this happen, the producers ask that the app would activate the microphone every minute to detect
if any music under their licence is being played. Third, music event promoters
are interested: by storing the IP addresses that are frequented by the mobile
device, on which the app is installed, and mapping these to latitude/longitude
coordinates, potential neighbourhoods for future events could be suggested.
Jenny loves the ideas, as a quick business case concludes that this could lead
to massive growth of her startup. When Jenny brings the ideas to her board,
she gets absolutely blasted. One investor openly questions her ability to lead
a data science related startup and threatens to fire her. As you can already
imagine, Jenny's story relates to the privacy of the app's users. This chapter
will start with detailing the importance of privacy and will discuss some concepts from the European General Data Protection Regulation (GDPR) that
can guide us in ethical data gathering. Next to these concepts, we'll cover
some privacy techniques, such as encryption, obfuscation, and differential
privacy, that look to reconcile privacy and the gathering and storing of personal data. Biased data science models often stem from bias in the data that is

[1] This is a fictitious story.

Data Science Ethics. David Martens, Oxford University Press.
© David Martens (2022). DOI: 10.1093/oso/9780192847263.003.0002

gathered, and will be discussed with several cautionary tales. Yet another data gathering practice that has important ethical considerations is human experimentation. We'll discuss this in a historical context and discuss the ethical issues of often occurring online experiments.

2.1 Privacy as a Human Right

2.1.1 The Importance of Privacy

Privacy is a likely first subject to come up when talking about data science ethics. This right not to be observed or disturbed has taken an important role in our current information age. As you move through your life, you send out a lot of personal information. When you buy something in the supermarket with a loyalty card, your name, and the products you bought are sent to some server. When you browse online with your smartphone, your mobile ID, IP address, and potentially latitude/longitude coordinates are sent to several servers. As you use the Internet, with every web page visited, post liked, payment sent out, or movie watched, personal information is sent along. Not only your name, address, and email address are sent, but also your PIN code and credit card number. These transactions in themselves are already private, but they can also be used to make quite accurate predictions about your political preference, product interest, and even personality (see e.g. [246, 356, 362, 365, 164]). Hence, such information should only be visible to and used by the specific entity you're transacting with, and only for the purposes you agree with.

In the Silicon Valley episode 'Facial Recognition'[2] COO Jared makes an analogy to the Manure Crisis of 1894 during an interview with Bloomberg TV:

Jared: I'm sure you're aware of the Great London Horse Manure Crisis of 1894.
Emily Chang: *I'm afraid I'm not.*
Jared: In the 1890s, the Industrial Revolution had people flocking to the city, and more people equals more horses, and more horses equals more manure. And it was predicted that by the middle of the next century, there would be nine feet of manure covering the streets. But what no one saw coming, was a new technology that would completely obliterate those concerns. The car. Over night, the manure problem vanished. And the Internet, as we currently know it, is rife with, uh, identity theft, and spam and hacking. So, it's manure ...

---

[2] https://www.imdb.com/title/tt7864446/

This manure is all related to privacy being violated. Jared continues to explain that their 'decentralized Internet' is the technology that will be as significant as the car. In the real world, there are other solutions for the manure problem, which we'll discuss in this book: awareness of the importance of privacy and what it means, legal frameworks like the European GDPR, and technology to facilitate privacy.

Several international conventions have stated that privacy is a human right, such as the 1953 European Convention on Human Rights (Article 8):

Right to respect for private and family life

1. Everyone has the right to respect for his private and family life, his home and his correspondence.
2. There shall be no interference by a public authority with the exercise of this right except such as is in accordance with the law and is necessary in a democratic society in the interests of national security, public safety or the economic well-being of the country, for the prevention of disorder or crime, for the protection of health or morals, or for the protection of the rights and freedoms of others.

The United Nations' 1984 Universal Declaration of Human Right (Article 12) states:

No one shall be subjected to arbitrary interference with his privacy, family, home or correspondence, nor to attacks upon his honour and reputation. Everyone has the right to the protection of the law against such interference or attacks.

A notorious 2013 legal case related to this fundamental human right is that of editor Ramon Zakharov versus Russia. In 1995 the SORM system (System for Operative Investigative Activities) was installed, which allowed for the interception of telephone communications in Russia, and required telecommunications providers to install the monitoring hardware [282]. Mr Zakharov alleged that the system violated his right to privacy and that he did not have any effective remedy [132]. After the Russian courts dismissed his case, as Mr Zakharov could not prove he was a victim of such an interception himself, the European Court took up the case [183]. This court ruled that there were indications of 'the existence of arbitrary and abusive surveillance practices, which appear to be due to the inadequate safeguards provided by law'. It further ruled that because of these shortcomings, 'the Court finds that Russian law ... is incapable of keeping the interference to what is necessary in a democratic society. There has accordingly been a violation of Article 8 of the Convention' [132].

The potential for abuse of such surveillance is not limited to governments, also large (or even small) companies with access to massive personal data on a large portion of the population, face this risk. The Cambridge Analytica provides an illustration. The political consulting firm reportedly harvested private information from Facebook of over 80 million users without their permission, which was then used for (targeted) political advertising [176, 426, 199]. For the individuals, their public profile, page likes, birthday, and current city were likely shared with Cambridge Analytica [199]. The data was reportedly obtained through an app, thisisyourdigitallife, built by an academic researcher. Hundreds of thousands of users were paid to take a personality test and provide their Facebook profile data. When uploading its data, also data on the user's Facebook friends were sent. The harvested profiles were consequently reported to be used in several election campaigns, most notably in the US [426, 199] and the UK [176]. Even though the Facebook platform policy did not allow this data to be sold or used for advertising (only to improve the user experience in the app), the damage was done. Facebook removed the app, suspended Cambridge Analytica, and required certification that the data had been destroyed [176]. Although it was certified to Facebook that 'the data was destroyed', copies of the data reportedly remained beyond Facebook's control [380].

2.1.2 You Have Zero Privacy Anyway. Get Over It

Some of you might think by now 'Et alors?'[3], why should I care? As the CEO of Sun Microsystems Scott McNealy famously said in 1999: 'You have zero privacy anyway. Get over it.' [411] This statement, made more than 8 years before the first iPhone was launched, can still ignite intense discussions. When I ask my business students to vote between *Privacy is a human right* and *You have zero privacy anyway. Get over it*, some still choose the latter. One often used argument is: I don't care about privacy because I have nothing to hide. Let's dig into this argument a bit deeper.

---

[3] When the French magazine *Paris Match* revealed that the (at that time) French president François Mitterrand had an extramarital daughter, he simply reacted with the now infamous words 'Et alors?', meaning 'So what?' [261].

Nothing to hide to whom? I believe the students mean that they don't mind that companies like Google or Facebook have access to their data. But even so, they of course have things to hide: from their parents, from their (ex)partners, from their professor; how much money they inherited; if and what evil thoughts they once had; what their body looks like; with whom and how often they have sex; the text messages they have sent throughout the years, and so on. Asking the students if they'd be willing to provide access to all messages ever sent on Facebook or WhatsApp to their partner (let alone their fellow students), few respond affirmatively. How would you feel about giving access to all your messages to your boss or colleagues? Or broader: putting all texts on a public website, accessible to all? Can also a webcam be installed in your house (including your bathroom and toilet), with a live feed on the same website? Of course not. Even if you would not mind, any reasonable person would agree that most people would want such data to remain private. The potentially disastrous consequences of sharing all internet browsing history is a thought experience brought out with humour in the *South Park* episode 'Fort Collins'.[4]

The argument of having nothing to hide is also used to support government surveillance. This reverses the legal argument. The presumption of innocence is still a legal principle international human right under the UN's Universal Declaration of Human Rights, Article 11. The original Latin expression was: '*ei incumbit probatio qui dicit, non qui negat*', or translated: 'the burden of proof is on the one who declares, not on the one who denies'. Edward Snowden frames it as follows: 'Arguing that you don't care about the right to privacy because you have nothing to hide is no different than saying you don't care about free speech because you have nothing to say' [407]. In the wake of terrible events, such as terrorism or other violent crimes, often calls for more surveillance are made. The right trade-off between privacy and safety is hard to define and is a culturally and regionally dependent question. Yet, the impact of surveillance on human rights and its long-term effects are important to be considered, a point we'll revisit in Section 2.4 on government backdoors.

Facial recognition software has become an important technology within this discussion. Although such technology to identify a person from an image or video has been around for years, the improved speed and accuracy of the detection algorithms, as well as the increased use in applications such as authentication on smartphones, has made facial recognition an important data science technology when it comes to the threat of mass surveillance. This discussion will be further explored in Chapter 3.

---

[4] https://www.imdb.com/title/tt5218492/

Now, what are good privacy practices? What concepts should I embrace? Although ethics go beyond the legal aspects on this matter, regulations such as the European GDPR can provide guidelines to these open questions.

2.2 Regulations

2.2.1 GDPR

The General Data Protection Regulation[5] (GDPR) is a European law that came into effect on 25 May 2018, covering privacy and data protection of European citizens [335]. Also non-European companies that process data on European citizens have to comply. The goal of the regulation is to bring European laws up to speed with new advances in processing personal data, and provide harmonization of the laws in European countries. Some see GDPR as one of the world's most robust data protection rules [343], and includes fines of up to 20 million Euro or 4% of the company's turnover. Even if you do not fall under the GDPR regulation, it provides interesting concepts and guidelines for data science ethics[6].

Data protection concepts
The first concept to investigate is: what exactly is personal data, and when is data anonymized? GDPR defines personal data as follows: 'any information relating to an identified or identifiable natural person ("data subject"); an identifiable natural person is one who can be identified, directly or indirectly, in particular by reference to an identifier such as a name, an identification number, location data, an online identifier or to one or more factors specific to the physical, physiological, genetic, mental, economic, cultural or social identity of that natural person' [335]. The inverse of personal data is likely anonymous data. Interestingly, the GDPR text does not mention the word anonymization or anonymous. So if personal data is data that can be brought back to an individual, anonymous data is data that cannot be brought back to an individual. The definition of pseudonymization that GDPR offers follows those lines: 'pseudonymisation means the processing of personal data in such a manner that the personal data can no longer be attributed to a specific data subject

---

[5] A European regulation is a binding legislative act, which must be applied in its entirety across the European Union. A directive on the other hand is a legislative act that sets out a goal, but the individual countries devise their own laws to reach these goals.

[6] Note that the point of this section is not to provide a full legal discussion on the law; for legal advice consult a legal textbook or the full GDPR text [335].

without the use of additional information, provided that such additional information is kept separately and is subject to technical and organisational measures to ensure that the personal data are not attributed to an identified or identifiable natural person' [335]. An example of pseudonymized data is encrypting personal identifiers such as the name and social security identifier. Transforming this back to an identifiable person is quite easy (when having access to the encryption key); hence it is not anonymous. Anonymized data do not fall under the GDPR legislation, but have a heavy burden to carry: there shouldn't be a method that allows to transform the data back to an individual. This is not an easy task, as the re-identification cases of Section 3.2 will reveal.

But is anonymous really about personal identifiers, and what constitutes a personal identifier? Any variable that allows to recognize the same person in and across multiple datasets should be regarded as a personal identifier [316], and hence makes the data pseudononymous instead of anonymous. A cookie is a random string that is assigned to your browser and does exactly that, it allows advertisers and ad tech companies to identify you across different websites[7] and locations. A US social security number is also a random number [408], yet currently regarded as personal information because it allows to link data across silos [38].

Barocas and Nissenbaum make the point that the value of anonymity is not in being nameless but rather in not being 'reachable: the possibility of knocking on your door, hauling you out of bed, calling your phone number, threatening you with sanction, holding you accountable – with or without access to identifying information' [38]. These days there is so much data out there which can be used to infer many things about you. Think for example of how what you like on Facebook can be used to predict a wide range of personality traits, such as IQ, political, and sexual preference [246], and even your creditworthiness [104]. We can now act on individuals in exactly the way that anonymity was supposed to protect against: a data scientist does not need to know your name or social security number to predict what movies you like to watch, political party you might vote for, events you might be interested in, or what your sexual preference is. So even if we work with anonymous or pseudononymous data, privacy remains an issue to actively think about. Do I want to show an ad to someone who I predict is likely pregnant, even though her name, address, or any typical personal identifier are not used, just an in-house identifier, associated with a loyalty card for example? This is what a large US retail chain

---

[7] Third party cookies allow users to be identified across websites, first party cookies only within a website.

was confronted with, when they sent out offers for baby products to predicted-to-be-pregnant customers, including a teenage girl [118, 206]. We'll return to this example in Section 3.4.

This brings us to the point of the privacy of the model subject. Privacy is almost exclusively used to think about the data subject. Assuming the data is obtained in an ethical (fair, transparent, and accountable) manner, we should also consider how the model is applied to the data of potentially other persons. In the previous example, you don't need to store the data of the customer of your supermarket (for longer than it takes to scan the products at a cashier) to provide her with a targeted coupon. If we predict political preference, we infer the political preference for a set of persons (the model subjects) that might not want to have this revealed, based on patterns in the data of people who did reveal this trait (the data subjects). Barocas and Nissenbaum emphasize that: 'anonymity is not an escape from the ethical debates that researchers should be having about their obligations not only to their data subjects, but also to others who might be affected by their studies for precisely the reasons they have chosen to anonymize their data subjects' [38]. Ethical data science is not a checklist to be followed; it is thinking about these fundamental principles, such as being fair to both the data subject and the model subject with regards to his or her privacy, thinking what techniques could be helpful, while potentially remembering related cautionary tales.

Lawful basis

Now that personal data has been addressed, when does GDPR allow us to process such data? Article 6 of the regulation provides six legal grounds [335]:

1. unambiguous consent of the data subject,
2. to fulfill a contract to which the data subject is party,
3. compliance with a legal obligation,
4. protection of vital interests of the data subjects,
5. performance of a task carried out in the public interest,
6. legitimate interest (subject to a balancing act between the data subject's rights and the interests of the controller).

This list has a couple of interesting concepts. First of all: *unambigous consent*. According to the GDPR text, this consent needs to be freely given, specific, in-formed, and unambiguous. As pointed out by Barocas and Nissenbaum [38], this principle assumes that the person providing consent understands how their consent plays out. But the task of explaining what data is used, how and for what purposes, is a daunting task. How can you provide the data subject with notice in an understandable format, that he or she is willing to read, and obtain consent. Online privacy policies seem to be the way that websites have dealt with this issue, although there is ample evidence that this is barely read [32, 38]. When was the last time you read the cookie or privacy policy of your favourite news website or social media platform, let alone of each website you visit? Sometimes an effort is made to write these policies in very simple to understand wordings, and easy to tick boxes, but that inherently will lead to a loss of information. A concept which Nissenbaum refers to as the 'transparency paradox' [327]. Informing and obtaining consent is important, though far from easy. Ethical thinking about what the user should need to know and consent to is an important part of this process.

Even without consent, GDPR has other grounds under which we can process personal data—for example, to notify law enforcement about a criminal offence that has been committed, based on access to personal data. The open-ended nature of the last provision, on *legitimate interest*, raises important questions [21]. This implies you can process personal data in order to carry out tasks related to your business activities. You still must inform the individuals about this processing though. A couple of examples: a travel company can apply a recommendation system to improve the online user's experience, by recommending other travel locations likely to be of interest to the user. This personalization does not require explicit consent and can be done on the grounds of legitimate interest [21]. Also covered are the use of personal data for direct marketing purpose, for example sending out a direct mail from your charity to existing supporters to notify them of upcoming events, or to prevent fraud, or to ensure security of your IT systems.

But where is the line in requiring explicit informed consent and legitimate interest? In an opinion paper by a European Data Protection Working Party, several scenarios for a pizza place are provided to discuss this [20]. The first scenario is the pizza delivery service sending out coupons by postal services, based on previous deliveries. Here an opt-out option is suggested, without the need for informed consent. The second scenario involves advanced targeted online and offline advertising, where the pizza ordering data is combined with other data, such as the local supermarket's dataset. When the data subject would move to another location, the supermarket bill might increase due to a change in the data. Although the data and context are rather innocent in nature, due to the scale and financial impact, informed consent is needed. In the third scenario the pizza place would sell its data to an insurance company to adapt health insurance premiums. The insurance company might argue that there is a legitimate interest in assessing health risks and risk-based pricing, and therefore wants the pizza buying behaviour data. However, a reasonable person would unlikely have expected that her pizza consumption behaviour would be used for calculating her premium. Given the sensitive nature of the data and the large impact of the data science, the rights of the data subject override the legitimate interests of the health insurance company. These scenarios illustrate that this balancing act explicitly relies on what a *reasonable person* would find acceptable and what the potential *impact* of the data science practices are. Ethical data science is all about this balancing act: what data can I use, for what purpose, and how should I go about it?

Principles relating to processing personal data
Article 5 of GDPR [335] can be seen as the heart of the regulation[8]. It provides good practices when considering privacy and gives insight into the spirit of the law. Figure 2.1 summarizes the first paragraph of the article. The second paragraph, Article 5.2, covers the accountability criterion, stating: 'The controller shall be responsible for, and be able to demonstrate compliance with, paragraph 1.' The six principles are crucial to keep in mind when gathering (and processing) personal data. As GDPR is a legal piece of work, these principles

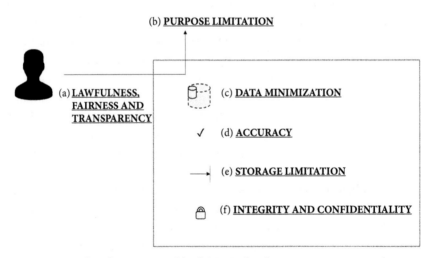

Fig. 2.1 Article 5 (1st paragraph) of GDPR detailing important principles relating to processing of personal data.

---

[8] https://ico.org.uk/for-organisations/guide-to-data-protection/guide-to-the-general-data-protection-regulation-gdpr/principles/

are made more concrete in the following paragraphs, with fines that have been imposed for violation of these principles.

5.1 (a) Whenever personal data is processed, this should be '(a) processed lawfully, fairly and in a transparent manner in relation to the data subject'. Definitely on the transparency requirement there have been some infringements. An interesting case is that of the Spanish national football league (La Liga), which was fined under this article [294]. They reportedly created an app where—in their fight against piracy—every minute the device's microphone was used to detect if the user was watching a football match. If so, the GPS location of the device was accessed as well to determine whether the venue, for example restaurant or bar, was showing the match legally. The technology used was reportedly similar to that of Shazam (used to identify music), and transformed the acoustic footprint into a hash. As we'll see later on in this chapter, hash functions are one-way, meaning that it is very difficult to go from the hash back to the original recording [294].

5.1 (b) The second item points to the importance of making the purpose of the data processing specified, explicit, and legitimate: personal data shall be 'collected for specified, explicit and legitimate purposes and not further processed in a manner that is incompatible with those purposes; further processing for archiving purposes in the public interest, scientific or historical research purposes or statistical purposes shall, in accordance with Article 89(1), not be considered to be incompatible with the initial purpose'. A simple thought experiment to illustrate: suppose you're the mayor of a rather small municipality. You're in contact by email with citizens who reached out to you through their architect with respect to an urban planning project. Elections are coming up, and you'd like to send out a political campaign email to these citizens. Should you? Well, a Belgian mayor did something very similar, and was charged a 2,000 Euro fine by the Belgian Data Protection Authority for non-compliance with the purpose limitation principle: data gathered for public services should not be used for personal election campaigns [358].

5.1 (c) The data should be 'adequate, relevant and limited to what is necessary in relation to the purposes for which they are processed'. A European payment services provider was fined for non-compliance with GDPR's article 5 (as well as other articles). It was determined that the company collected more personal data than it indicated as necessary for effecting a payment initiated by the payer (such as amounts and dates of other payments) [134].

5.1 (d) The data should be 'accurate and, where necessary, kept up to date; every reasonable step must be taken to ensure that personal data that are inaccurate, having regard to the purposes for which they are processed, are erased or rectified without delay '. Once it becomes certain that certain data points are inaccurate, these must be corrected. A Hungarian bank was fined under this article [33]. At the time of signing up a certain customer, a wrong telephone number was reportedly provided to the bank. Because of this, the bank mistakenly sent text messages about a customer's credit card debt to this wrong telephone number. When it became clear that this was a wrong phone number, the number was not erased and still text messages were sent to this person who wasn't a customer of the bank. A fine of around 1,600 Euro was reportedly imposed by the Hungarian Data Protection Authority [33].

5.1 (e) The gathered data should be 'kept in a form which permits identification of data subjects for no longer than is necessary for the purposes for which the personal data are processed; personal data may be stored for longer periods insofar as the personal data will be processed solely for archiving purposes in the public interest, scientific or historical research purposes or statistical purposes in accordance with Article 89(1) subject to implementation of the appropriate technical and organisational measures required by this Regulation in order to safeguard the rights and freedoms of the data subject'. A Danish taxi company was fined for keeping data on their customers for too long [133, 85]. Even though the name and addresses were removed after two years—as also stated in their data retention policy—the customers' telephone numbers were kept for another four years (as this was supposedly used as a key for their database). The Danish Data Protection Authority recommended a fine close to 160,000 Euro [85].

5.1 (f) Finally, the data should be 'processed in a manner that ensures appropriate security of the personal data, including protection against unauthorised or unlawful processing and against accidental loss, destruction or damage, using appropriate technical or organisational measures'. In a Portuguese hospital, it was found that some non-medical hospital staff, including social workers, had access to patients' data through false doctor profiles. There were a reported total of 985 registered users with the profile 'doctor' in the system, while only 296 doctors were employed [254]. Because of the lack of appropriate security measures, the confidentiality of the data was not sufficiently protected, as doctors, regardless of their speciality, and hospital staff had unrestricted access to patients' data. A fine totalling 400,000 Euro was imposed by the Portuguese Data Protection Authority for violations under this, and other, articles [309, 254].

Discussion 1

Consider the following scenario, inspired from a real case from Google Home [401]. Due to a leak from a Google subcontractor, Google Home recordings of Belgian persons were revealed to a journalist [197]. Google Home is a personal assistant where based on your spoken question, answers are provided. Different languages pose challenges in the performance of such personal assistants. Google stated that about 0.2% of all 'audio snippets' are transcribed by contractors to understand better the language difference, and which are not associated with user accounts [401]. The recordings reveal very private information for some users, including health related issues, and in some cases, Google Home started recording without the user intending to do so—for example because it wrongly detected 'Okay Google', which is the phrase used to activate Google Home [197]. As a data scientist, this practice of transcribing audio snippets makes a lot of sense, in order to improve the model by getting more ground truth to train the language-specific model. How about these ethical issues:

1. Would other data sources of speech be as good to do so? What problems would and would not be solved if other sources were used?
2. Would you make the process of transcribing samples more transparent and how should this be done?
3. What would other uses for this data be? Would these other purposes be aligned with the initial purpose?
4. Would the full conversations need to be stored and used? Or can you think of intelligent ways to limit the data such that transcribing these would still improve the model, while not providing semantics on conversations?
5. How would you ensure the transcriptions are accurate? What problems might arise in obtaining accurate transcriptions?
6. How long would you keep this data?
7. How would you divide the audio snippets among transcribers, in order to deal better with the data protection issue?
8. How would you ensure proper security measures so as to avoid leaks of such audio snippets to the general public?

2.2.2 Public Data Is Not Free-to-Copy Data

An often reoccurring misconception of data scientists is that public data is free-to-copy data. Data of public Facebook pages or online news articles, for example, are publicly available but are therefore not necessarily free to copy in your own database. Often businesses and startups have an idea on how to leverage an existing public dataset, but one should proceed carefully when gathering such data. Two principles are behind this warning: database rights and policies.

Database rights are a recognition of the investment needed to build up a database, and do not allow someone to copy (substantial parts of) the database without consent of the database owner [334]. Europe and the UK are regions with such a legal protection. In the European legislation, a database is defined as 'a collection of independent works, data or other materials which are arranged in a systematic or methodical way and are individually accessible by electronic or other means' [334]. This is quite a broad definition and therefore also includes mailing lists and telephone directories. These rights lasts for 15 years [334]. So even when the content of a database is not copyright protected, still the set of all items becomes protected when substantial investments were made to create this database. You are of course allowed to query the database, if it is public, but copying substantial parts is not allowed. Interestingly, article 50 foresees exceptions when the extraction is intended for private purposes, or for teaching or scientific research [334], though not for commercial purposes.

Next to database rights, often policies of the companies making certain data publicly available do not allow for the copying of it. Consider Facebook public pages, such as the one from the University of Antwerp. The idea that one can simply scrape (extract) all the content from such public pages, whether it is done manually or through some automated crawler, is not allowed: large websites will have a robots.txt file which tells a crawler (bot) what pages it is allowed to request. The very first line of www.facebook.com/robots.txt (as downloaded in early 2021) states in a comment that 'Crawling Facebook is prohibited unless you have express written permission.' Next it lists which webpages and directories are *not* allowed to be visited by specifically mentioned crawlers such as baiduspider, Bingbot and Googlebot. For all bots that are not listed in the robots.txt file, the final line prohibits the crawling of any page:

```
User-agent: *
Disallow: /
```

So whenever you are considering to scrape a website, have a look at the robots.txt file. Still scraping the website is at least unethical, and could also lead to legal headaches.

In such cases, most companies these days offer an API (Application Program Interface) to extract data from their platform. Both Facebook and Twitter for example offer such an API.[9] These APIs make it easy to retrieve data, in an ethical (and legal) manner. They do come with restrictions. Twitter provides both public and premium APIs, which vary in the number of tweets and number of days and years you can go back to retrieve tweets. And in the wake of the Cambridge Analytica story, Facebook became very strict in who to provide access to its API, even for academic researchers [42].

If you still decide to simply crawl such pages and copy them, remember that you are likely also storing personal data, such as the comments on public Facebook pages with names. That in itself brings along other ethical issues, and even legal ones when this personal data includes data from European citizens (cf. GDPR).

2.3 Privacy Mechanisms

2.3.1 Encryption

Encryption is probably the most important underlying method for data protection. The basic notion of encryption is that a message or information is encoded in such a way that only authorized persons can access it. In modern day society, it is a crucial tool to ensure that personal data is stored and communicated securely.

The procedure for encoding a message is called a cipher. Ciphers have been around for thousands of years: the Roman army for example used the Caesar shift cipher, a simple encryption technique where a letter is replaced by a letter that is some specific number of letters further along the alphabet. A three right shift would encrypt 'ETHICS' into 'HWKLFV'. The cipher is named after Julius Caesar who reportedly used such a three right shift cipher for his personal communication [434, 399]. Spartans used Scytale [399], where paper or leather, one character wide, would be wrapped around a rod of certain diameter in a helical way. The message would be written on the paper alongside the rod. It would then be unwrapped and sent to the receiver. When the paper or leather would be wrapped around a rod of the same diameter at the

---

[9] https://developers.facebook.com/docs/ https://developer.twitter.com/en/docs/twitter-api

receiver's side, the message would re-appear. In ancient Greece, encryption existed in the form of writing a message on a shaved head and letting the hair grow back before sending the messenger to the receiver (who would shave the head again) [399]. A well-known encryption device used by the Germans in World War II was Enigma [399, 301]. Breaking this code was famously worked on by among others Alan Turing and his team (as dramatically depicted in the movie *The Imitation Game*). The electro-mechanical machine had a set of rotors and plugs which defined the state of the machine. Each time a letter was typed on the machine, the machine output another letter. At the same time the state of the machine changed. So the same letter being input twice would (mostly) lead to two different letters being output. Both the sender (a U-boat submarine, for example) and the receiver (German headquarters) would need to start with the same state on their respective Enigma machines. By typing in the encoded letters on the Enigma machine at the receiver's end, the original message would come out. The initial state or machine configuration was noted on a secret page sheet (of a codebook) that both parties had a copy of, and would change every day. There were about $3 \cdot 10^{114}$ possible configurations [301]. Quite a lot, considering that there are only an estimated 10^{80} atoms in the universe. So trying all possible configurations on intercepted messages would not be possible. Additionally, the operator would choose a random initial word to start with, which would be used for further encoding of that message. Using some fascinating cryptographical insights, including known associations between encrypted and plaintext messages, discovered patterns, and the universal Turing machine, Engima was broken [399].

Symmetric encryption

These are all examples of what is called *symmetric* encryption, where one key is used for both encryption and decryption. This key is secret and shared only between the sender and receiver. Figure 2.2 illustrates the mechanism where sender Alice is sending the message 'Hello Bob' to receiver Bob. Using their shared key, the plain text is transformed into a cipher text. The three right shift cipher of Julius Caesar would, for example, encrypt this to 'Khoor Ere'. This cipher text is sent to Bob, who uses the same key to decrypt the message. For the Caesar cipher this corresponds to applying a three *left* shift operator. An eavesdropping person, Eve, can only see the cipher text, which makes little sense without the key. The attentive reader will have spotted the weakness of such a cipher, being the frequency of letters [224]. Knowing that the letter 'e' is the most frequent letter in the English vocabulary, looking for the most frequently used letter in all cipher texts reveals the encrypted version of the

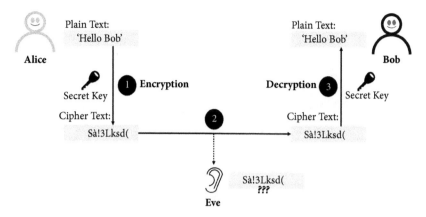

Fig. 2.2 Symmetric encryption.

letter 'e'. Additionally, typical words or phrases that are used to start or end a communication, for example 'Dear' or 'Sincerely yours', can easily be used to detect their encrypted version. Finally, a brute force attack would simply try all possible shift operators and quickly find the secret key.

DES (Data Encryption Standard) is one of the first major standards in modern symmetric key encryption, developed by IBM [399]. Although IBM reportedly intended to release DES with a key of 128 bits, the US National Security Agency (NSA) supposedly convinced IBM to release DES with only 56 bits, as it would be easier to decrypt intercepted messages [78, 399]. Even though with 56 bits more than $7 \cdot 10^{16} (= 2^{56})$ possible combinations are possible, from the beginning the relative small key size was considered a major flaw of DES [78, 111]. A simple brute-force attack would allow the secret key to be found. Its replacement, AES (Advanced Encryption Standard), developed by the Belgians Vincent Rijmen and Joan Daemen in 1998, uses 128, 192, or 256 bit keys and became the new standard in the late 1990s. As a 128 bit key already generates $3 \cdot 10^{38}$ key combinations, it is deemed quite secure [453].

There are two major challenges with respect to keys: how to share the key, and how to manage the keys. For symmetric encryption simply sending the key as plain text is dangerous, as Eve would then be able to decrypt all subsequent communication. So there is a need for additional communication overhead (see below), or the key needs to be exchanged by meeting in person. For the Enigma machine, the machine and initial configurations for each date would be provided at the start of the mission. To communicate in this way with hundreds of persons, you'd need hundreds of secret keys, share them in some

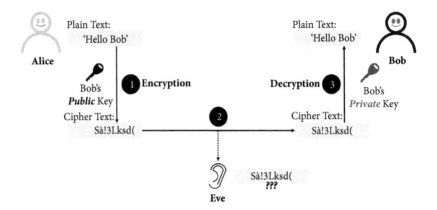

Fig. 2.3 Symmetric encryption.

secure manner initially, and then manage all the keys. As you need to share the key before communicating, it would also mean that for online computers you need to share keys with every computer that you might communicate with. If u users or computers need to be able to communicate with one another, this requires $(u - 1) + (u - 2) + \ldots + 1 = \frac{u \cdot (u-1)}{2}$ keys.

Asymmetric encryption
Asymmetric encryption is a relatively new approach that avoids the challenging problem of distributing a secret key. Instead of using a secret key, there are now two keys: a public one and a private one. The public key is revealed to the world, while the private key is kept secret. When Alice sends a message to Bob (see Figure 2.3), Alice will use Bob's public key to encrypt the message and send this to Bob. Only Bob can decrypt the message as it requires the private key of Bob, available only to Bob. Let's discuss this in more detail with the popular RSA algorithm [377].

The RSA algorithm is based on the fact that multiplying two large numbers is fast and easy; however, decomposing them into prime numbers, prime factorization, is very difficult. Try for yourself to compute $19 \cdot 13$, not so hard, right? But now try to decompose 391 (or 826.513 if you're a math wizz), knowing that it is a multiplication of two prime numbers. This is much more difficult to do although the prime numbers are of the same order as with the first task. When the numbers are sufficiently large, no efficient non-quantum (we'll get back to this) integer factorization algorithm is known [399]. So if the numbers are sufficiently large, currently the prime factorization is practically infeasible to calculate. More formally, if p and q are large prime numbers, calculating

$n = p \cdot q$ is easy, but given n, finding p and q is very difficult (practically in-feasible). RSA uses this idea, by making n public, and keeping p and q private. A message m would be encrypted to c as follows by Alice, using the public key n and some number e that is also made public (chosen such that e is prime relative to $(p-1) \cdot (q-1)$):

$$c = m^e \bmod \ n. \tag{2.1}$$

This is decrypted by Bob, knowing the private key p and q, but also e from his own public key. The integer d is a function of these: $f(p, q, e)$, and hence also part of the private key (for those interested, d is chosen such that $d \cdot e \bmod [(p-1) \cdot (q-1)] = 1$).

$$m = c^d \bmod \ n. \tag{2.2}$$

Let's use a simple example to illustrate these calculations:[10] we generate two prime numbers, $p = 7$ and $q = 3$. The multiplication becomes $n = 7 \cdot 3 = 21$. Suppose we want to send a message which consists only of the letter 'l'. We transform this letter to the number 12, as it is the 12th letter in the alphabet. We also need to define e such that e is co-prime to $(p-1) \cdot (q-1) = 12$, meaning it has no common denominators other than 1. An $e = 5$ satisfies this. The numbers e and n make part of Bob's public key. Someone who wants to send the message 'l' to Bob will calculate:

$$c = m^e \bmod(n) \tag{2.3}$$
$$= 12^5 \bmod(21) \tag{2.4}$$
$$= 248,832 \bmod(21) \tag{2.5}$$
$$= 3. \tag{2.6}$$

Next to p and q, Bob also has d as part of its private key. This number d must satisfy:

$$d \cdot 5 = k \cdot 6 \cdot 2 + 1 \tag{2.7}$$
$$= k \cdot 12 + 1. \tag{2.8}$$

---

[10] In reality these would be very large prime numbers, and as we want to do the calculations ourselves, we keep it very small.

For $k = 2$ we get $d = 5$. When Bob receives the encrypted message $c = 3$, it will decrypt this as:

$$m = c^d \bmod(n) \tag{2.9}$$

$$= 3^5 \bmod(21) \tag{2.10}$$

$$= 243 \bmod(21) \tag{2.11}$$

$$= 12. \tag{2.12}$$

And the original message of $m = 12$ re-appears. The only way that this is possible is when the factorization of n into the two prime numbers p and q is known. Eavesdropping Eve will hence not be able to perform this calculation.

The RSA algorithm is named after its inventors, Rivest, Shamir, and Adleman, who applied for the patent in 1977 [377, 399]. It was later revealed that the initial idea of 'non-secret encryption', encryption without the need of a secret key, was proposed earlier by Ellis in 1970 at the UK communications security group, but this work was deemed classified at that time [127].

Asymmetric encryption has several advantages [399]. First, it immediately solves the challenge of securely sharing the key. Now only the public keys need to be shared, and these, by definition, do not need to remain secret. Additionally, it limits the number of keys required. If we want u users to be able to communicate with one another, we would need to share only u keys (and each user keeps its own private key), as opposed to the $\frac{u \cdot (u-1)}{2}$ keys needed with symmetric encryption. On the other hand, asymmetric encryption takes relatively more time than symmetric encryption.

A combination of both is often used online, where a client and a server need to communicate securely. The client encrypts a random number with the public key of the server and sends this encrypted message to the server. With its own private key, the server can decrypt the message and also knows the random number generated by the client. This random number is then the secret key that is used with symmetric encryption in all subsequent communication, until the end of the session. This approach is the basis of the Secure Socket Layer (SSL) and subsequent Transport Layer Security (TLS) protocols, now widely used online [399, 78]. The management of a public key infrastructure, including validating that a certain public key corresponds to a given entity, is often done through a third-party Certificate Authority (CA), such as Comodo[11] and Let's

---

[11] https://www.comodo.com/

Encrypt.[12] Whenever you visit a webpage starting with https, this type of encryption is working in the background. If you see http instead, the connection
might not be secure.

Encryption for data protection

Encryption is useful for a variety of tasks to protect personal data. First, as
much of the information that is being sent online is personal, be it a password,
credit card number, pin code, email address, or simply even the webpages you
are visiting, it is important to ensure that it is sent securely. If you have a website
or app, make sure you use the TLS protocol. It may even boost your ranking
in search engines [30]. Also for storing the data, encryption is of course very
important, definitely when data is stored on removable storage devices such as
USB storage devices.

But personal data is not only stored on laptops, PCs, smartphones, and USB
storage devices. You might be making a device for bikes, where personal data
on biking performance is being stored. Encrypting the data would be recommended in that case as well. Car makers too should consider encryption, as
personal data, such as address books, mobile app login information, and addresses used in your navigation system, are being stored in your car. In the
US, the Federal Trade Commission even came out with an advisory for consumers to clean out the personal information before selling their car [435]. An
interesting related car feature comes from Tesla, where the cars include a Valet
Mode where personal data is hidden (among other functions) when giving the
keys to a valet [139]. On the other hand, in 2019, it was reported by CNBC that
some data in Tesla cars was stored unencrypted, as security researchers were
able to extract unencrypted personal data from junked Tesla vehicles [139].
The extracted data included a video of what happened just before the crash
of the car, calendar entries with descriptions of appointments and email addresses of those invited [139]. Storing this data in the car might be justifiable
and valuable to both car makers and car owners; however, storing this data
encrypted, just as personal data would be stored on your smartphone or PC,
is recommended for data protection.

2.3.2 Hashing

Hashing is another very useful cryptographic function, which maps some input to a value called hash or digest [78]. The hash is always of the same length.

---

[12] https://letsencrypt.org/

Table 2.1 Encryption versus hashing.

	Encryption	Hashing
Function	two-way plaintext $\leftrightarrow$ ciphertext	one-way plaintext $\rightarrow$ message digest
Main goal	confidentiality	integrity
Application	need to know plaintext	don't need to know plaintext
Length	variable	fixed

What is important is that the hash function is a one-way function: it is easy to go from some input to the hash, but given the hash value it should be very hard to find the corresponding input. This is one of the differences with encryption (see Table 2.1), where we still want to be able to reverse from the cipher text back to the input text. The total space of all hash digests is typically smaller than the total space of possible inputs. This can lead to so-called hash collisions, where two different inputs are hashed to the same output. Larger output spaces will have less collisions.

The popular hashing algorithm MD5, named after Merkle and Damgård, takes any string as an input and outputs a 128 bit digest [78]. The more recent SHA-3 hash algorithm, released by the National Institute of Standards and Technology (NIST) in 2015 [329], outputs hashes of up to 512 bits. The developers are Bertonu, Daemen (remember him from also co-inventing the AES encryption standard), Peeters, and Van Assche.

A first interesting use is message fingerprinting. A message, such as an email, can be hashed onto a certain value. If anyone changes the message, one can easily verify that the hash of the message no longer matches the message digest (hash value). Hashing is also very useful to store personal information, such as passwords. Suppose you run an e-commerce website with a login module. A first, naive approach to storing the user name and password of your customers, is by storing exactly those two fields, as illustrated at the top of Figure 2.4. This has serious security flaws: as the password is sent over the Internet, if your website does not use TLS, the password is visible to eavesdropping Eve. Also, if your database is ever leaked, all your customers' passwords will become compromised, not to mention that potentially some employees of your company are able to snoop in the password list of your customers. This is where hash functions come in: instead of storing the password, hash the password at the client's side, send the hashed value to the server and store this value. As the function is one-way, it is very difficult to determine the actual password based on the stored hash value. A data leak would not cause that much harm any

User	Password
David Martens	123456
Jennifer Johnsen	123456
Latifa Jenkins	p@ssword5
...	

User	Hash
David Martens	e10adc3949ba59abbe56e057f20f883e
Jennifer Johnsen	e10adc3949ba59abbe56e057f20f883e
Latifa Jenkins	f3f092cd075b3e050451239611a9e1e9
...	

User	Salt	Hash
David Martens	t0mR1aoPdp	79e514abbfb414c5ae58b553fbb0ff00
Jennifer Johnsen	rmsP5dof9y	896cab0975703f599fe0f7491c90062b
Latifa Jenkins	8LyqGp4cPm	8a945d114467eee63485fba9aa0bbaf0
...		

Fig. 2.4 Different approaches to storing a password: storing the password itself (top), storing a hash of the password (middle), and storing the hash of a random salt concatenated to the password (bottom).

Popular passwords	Hash (MD5)
123456	e10adc3949ba59abbe56e057f20f883e
password	cc3a0280e4fc1415930899896574e118
123456789	6ca5ac1929f1d27498575d75abbeb4d1
12345678	25d55ad283aa400af464c76d713c07ad
12345	827ccb0eea8a706c4c34a16891f84e7b
111111	96e79218965eb72c92a549dd5a330112
1234567	fcea920f7412b5da7be0cf42b8c93759
sunshine	0571749e2ac330a7455809c6b0e7af90
qwerty	d8578edf8458ce06fbc5bb76a58c5ca4
iloveyou	f25a2fc72690b780b2a14e140ef6a9e0
princess	8afa847f50a716e64932d995c8e7435a
...	

Fig. 2.5 A hash table showing the MD5 hash of the 10 most popular passwords in 2018 [336].

more. However, because many people typically use the same password, very frequently occurring passwords can likely be identified. Figure 2.4 and 2.5 illustrate this: we first make a table listing the most common passwords and their corresponding hash value, known as a 'rainbow table' (Figure 2.5 lists the most frequently passwords in 2018, according to DataSplash [336]).[13]

If we now look at the hash value of David's password, we find it in the popular password table, and derive that the password will be 123456. Of course the number of possible passwords is huge, and such a hash table would only be feasible to build for a subset of possible (popular) passwords. A way to improve on this is to generate a random string for each user, called the salt, which is concatenated to the password. As shown at the bottom of Figure 2.4, now the hashed passwords of David and Jennifer are no longer the same, nor can they easily be derived from the constructed rainbow table. In case of a data leak, the salt of each user will likely be leaked as well, and a hacker could now create a rainbow table for each user (adding the salt of David at the beginning of all popular passwords and calculating the hash). But this now requires much more computation than when not including a salt.

This hashing of passwords is becoming a standard to follow, as illustrated by the GDPR fine that was imposed on Knuddels.de [250]. Knuddels is a German chat community which was hacked, in which, according to reports by *Der Spiegel*, over 800,000 email addresses and passwords were published on the Internet [250]. The hacking itself was not the reason for the fine; rather, the storing of passwords in plain text was the issue: 'By storing the passwords in clear text, the company knowingly violated its duty to ensure data security in the processing of personal data pursuant to Article 32 (1) (a) of the GDPR.' [267] The company was fined 20,000 Euro.[14]

Hashing can also be used to ensure that no personal data is copied throughout your data processing system. Datasets often get copied and downloaded on several devices, in different formats. If a user asks to have all of its data removed, it can be a very cumbersome task to ensure that the data is no longer stored on any device or location. Often you don't need to know the personal information, such as exact name or social security number, in most of your system and applications anyway. Figure 2.6 illustrates how hashing can help with this situation. Instead of using some personal information throughout your data processing systems, use a hashed value of the personal information (for example, the name and social security). Only keep the personal information, with its hashed value in one table. In all others, use the hashed value. So when Jennifer asks to remove all her data, simply removing the line in the first

---

[13] The top 25 list shows interesting cultural happenings, for example 'batman' made the list in 2014, 'superman' in 2011 and 2014, 'starwars' in 2015 and 2018, 'solo' in 2015 and 2016 and 'donald' in 2018 [463].
[14] The relatively small fine was attributed to the exemplary fashion in which Knuddels.de reported and reacted to the incident [348].

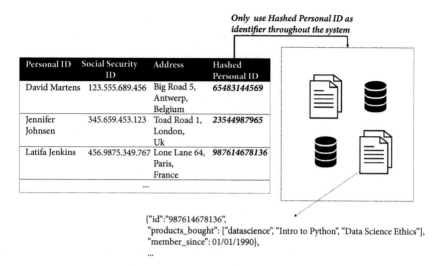

Only use Hashed Personal ID as identifier throughout the system

Personal ID	Social Security ID	Address	Hashed Personal ID
David Martens	123.555.689.456	Big Road 5, Antwerp, Belgium	65483144569
Jennifer Johnsen	345.659.453.123	Toad Road 1, London, Uk	23544987965
Latifa Jenkins	456.9875.349.767	Lone Lane 64, Paris, France	987614678136
		...	

{"id":"987614678136",
"products_bought": ["datascience", "Intro to Python", "Data Science Ethics"],
"member_since": 01/01/1990},
    ...

Fig. 2.6 Use of a hashed personal ID to ensure no personal data is stored in the system. Upon removal of the line in the hashed personal ID table, other data in the system (on servers, copies on laptops, etc.) will no longer contain or be linked to personal data of the customer.

table will do most of the work. Of course, you should attempt to remove even the data with the hashed value from all systems. But if some employee has a copy of data that you don't know about on Jennifer—what books she bought, for example, in a dataset that he copied to an Excel file to his laptop—he can no longer link this (easily) to the person Jennifer. If her name and address were stored in the Excel sheet, you would still (unknowingly) have personal data stored on her.

This is an example of what GDPR calls 'pseudonimization'. It's of course important to keep the hash table secure (for example encrypted and/or in another system) or even remove this altogether. Such pseudonimization can also be helpful for sharing data with researchers, though still confidentially. A large bank could for example share its pseudonomized payment data, by hashing account numbers and removing names for example, with a trusted university or researchers (after signing an NDA). It might not be anonymized, as it could be possible that someone is able to re-identify some of the customers in the data, yet the hashing already provides additional data protection safeguards.

Both encryption and hashing are crucial technical measures in the quest for data protection. Remember though that hackers are constantly looking for weaknesses in implementations and current standards in a ceaseless arms race. Personal data protection of 100% is only assured when no personal data is stored. One fascinating way in which encryption might be overpowered in the future is quantum computing. Due to its large foreseen impact in the future on encryption, a short introduction follows next.

2.3.3 Quantum Computing

Quantum computing allows us to make computations in a time that no conventional computer can by using the powers of quantum mechanics [325]. The unintuitive mechanics at a quantum level are used in near-magical science. The unit of information within a quantum computer is a *qubit* which can be both 0 and 1 at the same time. Physicist Erwin Schrödinger devised the following thought experiment to interpret quantum mechanics with everyday objects[15]: suppose we place a cat in a closed box, we name the cat Schrödinger's cat [325]. The box also contains a deadly poison, which has a 50% chance of being released and killing the cat.[16] Once we've placed the cat in the box, is it alive or dead? Well, with the effects of quantum mechanics in mind, it is both. Only when we open the box and observe the cat will it be either dead or alive. This near-philosophical thought experiment actually is what really happens at atomic levels.

A key phenomenon is called *superposition*, where a particle, such as a photon, is in two states at the same time. As soon as there is an observation of the state, the superposition collapses into one of the states. As strange as this may be, it has been validated by experiments, for example with a photon having different polarizations or the state of an electron orbiting a single atom [325]. A qubit uses this superposition property of basic particles such as a photon or the spin of an electron, so it can be both 0 and 1 at the same time,[17] as illustrated in Figure 2.7. Whereas one traditional bit can be only one of two number: 0 and 1, a qubit can be both at the same time. Two qubits can store four numbers, or better, can be in a quantum superposition of four states: 00, 01, 10, and 11. And more generally, n qubits can store 2^n states at the same time.

The second important property of qubits is *entanglement*. The observation of one qubit instantaneously also reveals the state of the entangled qubit, no matter the distance between the qubits. Referring back to Schrödinger's cat, suppose we now have two boxes, each with one cat, which are entangled so that if one dies, the other lives. As soon as one of the two boxes is observed (opened), the other instantaneously will take the other state, even if the other

---

[15] Be aware that some physicist might object to the simplification of quantum theory in this story.
[16] Do not try this at home as half of you will not like the outcome.
[17] More specifically, it can exist in a continuum of states between 0 and 1, until the qubit becomes observed [325]

box were at the other side of the galaxy. This unintuitive phenomenon is what Einstein called 'spooky action at a distance' [296]. Different experiments have shown this to be what indeed happens in reality. In a 2017 experiment, for example, entangled qubits were 1,200 kilometers apart (a Chinese satellite and a ground station) [474].

The key to quantum computing is to use these principles of superposition and entanglement to represent a problem and add a way to assess the answer. The quantum decoherence of the qubits is set up in such a way that only answers that pass the test survive the decoherence. Shor's algorithm leverages the massive distributed computations of quantum computers for factoring large numbers, and is considered as one of the major leaps forward in quantum computing [389]. The factoring of large numbers is exactly the basis of the popular RSA algorithm for asymmetric computation. So being able to do so would 'crack' such encryption.

But we're not there yet: in 2020, quantum computers operate with only a few dozen of qubits [107, 381], while it's estimated that to break RSA a quantum computer with thousands of qubits would be needed [107]. In 2019, Arvind Krishna, who was IBM's senior vice president of cloud and cognitive software at that time and would become CEO a year later, predicted that quantum computers would crack the encryption of that time within a decade [91].

Larger quantum computers would have a major impact on encryption and the security of our personal data. Companies like Google, Microsoft, and IBM have reportedly launched projects to create quantum computers, and governments as well are investing heavily in this technology [91]. However, obstacles still remain, such as having to avoid any thermal interference that would collapse the superposition, arguing for an environment close to zero Kelvin [107]. There is also some scepticism about the practicalities of scaling up quantum computers [231]. Still, quantum computing undoubtedly has fascinating and potentially massive implications when it comes to data protection.

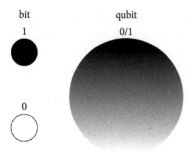

Fig. 2.7 A bit is either 0 or 1, while a qubit can be in a superposition of both 0 and 1 at the same time.

2.3.4 Obfuscation

Hiding personal, secret data is a key mechanism to deal with privacy. Encryption is one way to do so, by explicitly changing the secret data to an encrypted version that only authorized persons can read. End users, however, often have little impact on whether and how encryption is used. In the online world, our search queries, our web browsing data (which includes what webpages we visit), what ads we see and click on, from what type of devices, and sometimes even location data, is used in a complicated ad tech world, often by various players. Obfuscation is another method to hide your personal data, where end users deliberately add noise to the system [71], for example, by automatically generating a large number of search queries, so that your own queries are lost in this large volume. Or by installing a web browser plugin that automatically clicks on all ads, on each webpage you visit (combined with an ad blocker), so that ad tech companies no longer know what product or ad interaction was truly yours.[18] Formally, obfuscation is defined by Brunton and Nissenbaum as 'the deliberate addition of ambiguous, confusing, or misleading information to interfere with surveillance and data collection'. [71] Obfuscation differs from encryption methods in the sense that obfuscation does not explicitly hide your data, but rather implicitly, by generating many other data (the noise) so as to conceal the real data (the signal) in the vastness of the created data.

In their book, Brunton and Nissenbaum provide over 30 compelling cases that illustrate how obfuscation can be used [71], from planes in World War II releasing strips of black aluminium paper that fill radar screens with many signals, to babble tapes where dozens of recorded voices are played in order to hide your own conversation in potential recordings, and the swapping of supermarket loyalty cards, both physically and online. Also several interesting cases from the online world are discussed. The *TrackMeNot* software strategy was developed by the same authors in 2006 in the context of the US Department of Justice requesting Google to hand over search logs and the AOL re-identification case involving search queries (we'll come back to this in Chapter 3). The goal of the software is not to hide search queries but to obfuscate them by adding automatically generated search queries. These are generated in an intelligent manner, such that different users develop search queries from different initial lists of terms. In that way it becomes harder to detect whether a query is obfuscation (noise) or real (signal).

---

[18] This exists in the plugin **AdNauseam**: https://adnauseam.io/

Obfuscation doesn't always need a predefined technical system to add the noise. Humans can be the bringers of noise themselves as well: in November 2015, Brussels was in lockdown as the authorities were conducting raids on suspected terrorists. The police explicitly asked the public not to reveal police's actions and movements on Twitter as not to tip off the suspects. Citizens soon started flooding Twitter with images of cats with the tag #BrusselsLockdown [404]. Anyone who tried to find information on current police movements would have to scroll through thousands of (amusing) cat images. This is another beautiful example of creating 'a needle in a haystack' that obfuscation calls for.

But is obfuscation not an unethical practice in itself, as users intentionally lie, and might be considered to be free-riding in an ecosystem where services are provided for free in exchange for seeing ads? The Brussels Twitter cats example demonstrates that there are clear cases where obfuscation can be of great help to society. Brunton and Nissenbaum argue furthermore that it is warranted to counter information asymmetry, when data is collected in a context that is disproportionate and inappropriate [71].

As a business it is important to realize that this type of method might be used. The AdNauseam browser plug-in, for example, that functions in conjunction with uBlock Origin quietly clicks on all blocked ads. Ad tech companies tracking the ads will observe that the user has clicked on all ads, and hence is no longer able to detect what product or ad the user actually is interested in. Clicks are often the measure used to evaluate a campaign [101], and to define the cost of an ad. The CPC (cost per click) model will suffer greatly from this type of obfuscation, and some advertisers might see many clicks (costs) without the ads ever being shown. Such mechanisms can be detected by independent reporting of ad viewability and clicks, which likely will detect the uselessness of the data of these users. Reporting transparently when this is detected (not trying to charge excessive marketing budgets for bogus clicks) is of course another ethical implication.

2.3.5 Decentralized Differential Privacy

Adding noise is an important aspect of ensuring privacy [155], an approach that is used in differential privacy. The goal is to add noise such that the data can still be analysed, while ensuring (more) privacy. This noise can be added directly to the *data* itself, before it is being recorded, or to the *result* of the analysis. In the former decentralized (or local) approach, one does not trust the data analysis, while in the centralized approach, the data analyst is trusted but not the outside observer who is given the result [123]. The name differential privacy stems from its definition, which describes that the analyses with or without your data should not differ much [467]. We'll revisit the definition more formally in Section 4.1.1, when centralized differential privacy is introduced. For now, we'll focus on the local, decentralized version where noise is added when the data is being recorded.

Consider the case when a group of students of a 'Data Science Ethics' course is asked whether they like the course or not. Some students fear that if they respond that they disliked the course, the exam might be made more difficult for them or their grade might be lower. As the professor obtains and analyses the data, the students don't trust the data curator and hence require a decentralized differential privacy setup. So how will noise be added to the responses? A randomized response setup can then be used [458], where for each response, before it is being recorded, a coin is flipped. If it's heads, then the real answer is written; if it's tails, we flip the coin again. If that is heads, then the real answer is written, but if it's tails again, the inverse of the real answer is written. By doing so, we know that in 75% of the cases the answer is correct, and in 25% of the cases it is wrong. Anyone whose response is included in the dataset now has a plausible deniability, and can always claim that the opposite answer was given.

Such a setup is useful in any situation where sensitive data is being gathered and the data curator is not trusted. Think of surveys asking you about diseases you once had, your sexual and political preference, or whether you ever cheated on an exam or your partner. The procedure can be done by the respondent herself, or can be included in the software that records the answers (for example in an online survey). If the data were hacked, leaked or subpoenaed, one can simply deny that the recorded answer was the answer they gave [123].

This seems nice, but are the results still useful then? For example, if it is reported that 80% liked the course, this number is not the true number because of this addition of noise, right? Indeed, but we can calculate what the actual positive respondent rate was if enough respondents were included. Suppose p percentage responded positively, then how many would we observe after the coin flipping procedure? We know that in 3 out of 4 cases the actual response is recorded, and in 1 out of 4 the inverse is recorded. So the probability for a positive outcome is [123]: $\frac{3}{4} \cdot p + \frac{1}{4} \cdot (1 - p)$. For $p = 80\%$ we would observe a positive response rate of 65%. So for a large enough sample, we can confidently calculate back what the actual positive response rate would be, while providing plausible deniability to all respondents. This is a great way to calculate populations statistics.

The level of privacy that is guaranteed of course depends on the amount of noise added (in 25% of the cases in our example earlier in this section), but also on the number of times an analysis is conducted: the more an individual takes part in the same survey, the more certain one can become about the individual's response. This is what we will later, in Section 4.1.1, link to a 'privacy budget'.

Some large tech companies have started to use this procedure, as they realize that collecting user data statistics can help to improve security, find bugs and improve user experience, yet privacy should be guaranteed as much as possible while doing so. Google for example reportedly uses decentralized differential privacy to learn statistics about Chrome from individual users [128, 175], while Apple applies this technology to improve features as QuickType and emoji suggestions, and find Safari Energy Draining and Crashing Domains [15, 177].

Discussion 2

Suppose a mobile operating system (OS) wants to conduct an analysis on the frequency of occurrence of a new crying emoji in text messages. The OS sends as response whether on a given day, any of the text messages included the new crying emoji. They apply the decentralized differential privacy mechanism, where noise is added to the response on the mobile phone, before it is being sent to the server.

1. What could potentially be inferred about an individual if the actual response for that person would be revealed?
2. Suppose the mobile OS sends this information from your mobile phone daily. Would this increase the privacy risk, and how?
3. In the observed responses 30% of the texts have reportedly used the emoji. What would be the actual number if the procedure outlined in the previous paragraph was used, where in 75% of the cases the true answer is provided?
4. What is the risk if the actual answer is recorded in 99% instead of 75% of the cases?

2.4 Cautionary Tales: Backdoors and Messaging Encryption

2.4.1 Government backdoors

San Bernardino, California, 2 December 2015. A married couple of Pakistani descent enters a training event and Christmas party for about 80 employees and start shooting: 14 people are killed and 22 others are seriously injured [272]. After fleeing, they are shot by the police in a shootout. Upon discovery of the locked-out work phone of the male shooter, an iPhone 5C, the FBI stumbles upon the iPhone's security feature that would erase all data after 10 consecutive unsuccessful attempts to guess the passcode [424]. Apple is asked by the FBI to write software for the phone that would bypass the security feature so as to allow a brute force attack. Apple refuses [252]. This case is just one example of a long, ongoing demand of governments to have the ability to access data on mobile phones. The motivation is clear: after obtaining a warrant, law enforcement agencies are also allowed to invade your privacy by entering your home or searching your vehicle. So the line of thought is that they should also be able to access your phone.

Another example of government backdoors is the restriction that the US and its allies wanted on cryptography to resist decryption by such intelligence agencies as the NSA. In the 1990s, the longest encryption key size that was allowed for export without an individual license was 40 bits. So Netscape made two versions of its web browser: a US version with an encryption key of 128 bits, and an international one with a key size of 40 bits. Note that the 88 extra bits meant that the encryption was $300 \cdot 10^{24}$ times stronger [78]. The weaker version could be broken in a matter of days on a regular computer. At that time, Microsoft Internet Explorer was reportedly set to weak encryption by default, where a patch had to be loaded for 128 bit encryption [78]. As such, the government's demand actually led to worse encryption for many people. Since 2000, new US regulations simplified the export of cryptography, including allowing unrestricted key length restriction (after obtaining approval) [74].

Case closed? As strong encryption could be used now? Not really: in 2013, Edward Snowden revealed that the NSA had the ability to read encrypted communications. Methods reported by *The Guardian* included 'covert measures to ensure NSA control over setting of international encryption standards, the use of supercomputers to break encryption with "brute force", and – the most closely guarded secret of all – collaboration with technology companies and internet service providers themselves' [34]. In the years that followed, Apple developed new encryption methods for its iOS based devices, so that Apple could no longer comply with the government's request to extract data from iOS devices [386].

Even more recently, governments pressure mobile phone companies and telecommunication companies for backdoors in order to obtain access to personal data [135, 145]. The Five Eyes intelligence alliance, comprising the governments of the United States, United Kingdom, Australia, Canada, and New Zealand, has issued a 'Statement of Principles on Access to Evidence and Encryption' in 2018 which suggests that telco and tech companies will face strong opposition if they do not provide law enforcement and governments with backdoors for 'lawful access' to encrypted information of citizens [135].

2.4.2 Arguments for a Government Backdoors

This brings us to the main argument for government backdoors in encryption: privacy is not absolute. Just as governments and law enforcement can get access to your home, vehicle, and the like, when obtaining a warrant, they want to access your mobile phone and other electronic devices as well. The Five Eyes alliance states: 'that appropriate government authorities should be able to seek access to otherwise private information when a court or independent authority has authorized such access based on established legal standard' [145].

They agree that privacy is important and the crucial role of encryption in protecting those rights. Yet, the issue they address is the presence of 'challenges for nations in combatting serious crimes and threats to national and global security'. They want support from industry when they lawfully need access to personal devices. Under the first section 'Mutual responsibility' it is stated: 'Providers of information and communications technology and services – carriers, device manufacturers or over-the-top service providers – are subject to the law, which can include requirements to assist authorities to lawfully access data, including the content of communications.' [145]

The threat in the document of the Five Eyes alliance comes in the last sentence: 'Should governments continue to encounter impediments to lawful access to information necessary to aid the protection of the citizens of our countries, we may pursue technological, enforcement, legislative or other measures to achieve lawful access solutions.' [145]

The argument, privacy is not absolute, is quite powerful: few people would question the right of the police to enter a home after a warrant is lawfully obtained, or the similar right to search a vehicle or obtain phone records of a known terrorist. Lindsey Graham, a US senator, framed this discussion during a presidential candidate debate in 2015 as follows: 'Any system that would allow a terrorist to communicate with somebody in our country and we can't find out what they're saying is stupid.'[459]

During his tenure as FBI director, James Comey has often expressed his concerns about encryption in apps [225]. He warned that law enforcement officials were "going dark" as bad actors increasingly rely on encrypted messaging for their criminal acitvities, while there is a limited technical ability to intercept and access these communications and information pursuant court orders. In a speech in 2014 on the issue he states [225]: 'Encryption isn't just a technical feature; it's a marketing pitch. ... Sophisticated criminals will come to count on these means of evading detection. It's the equivalent of a closet that can't be opened. ... And my question is, at what cost?' He agrees that citizens should be sceptical of such power at governmental level, but that 'it's time that the post-Snowden pendulum be seen as having swung too far in one direction – in a direction of fear and mistrust.'

Not only politicians and law enforcements put forward this argument of privacy not being absolute. Also business men, such as Warren Buffett, agree. In an interview on CNBC in 2016 he argues: '[when the] AG [(Attorney General)] or head of FBI [is] willing to sign and go to a judge, saying we need this information, we need it now, I would trust that official to behave in a proper manner.' [252]

Business professor and author of the *New York Times* bestseller book *The Four: The Hidden DNA of Amazon, Apple, Facebook and Google* [160], Scott Galloway, agrees with this line of thinking. He illustrates the point with the case of Brittney Mills, a 29-year-old woman who was gunned down in 2015 in Louisiana (United States). At that time she was eight months pregnant. The baby did not survive either. Investigators found her password-protected iPhone and believed the identity of her murderer was likely to have been on the phone. Apple refused to comply with a search warrant to unlock her phone [125, 392]. 'Have we decided that the phone or specifically the iPhone warrants some divine exceptional rights not afforded to your other devices or your phone records or your bank accounts?' Galloway argues [392].

A more forceful version of this argument of a need for balance between privacy and security comes from Donald Trump who in response to the FBI/Apple discussion suggested in 2016 to: 'Boycott Apple until such time as they give that security number.'[19] [113]. Actually, both 2016 presidential

---

[19] No 'security number' was asked for; rather, software that would disable certain iPhone security features so as to allow a brute force attack.

candidates, Hillary Clinton and Donald Trump, sided with law enforcement authorities on giving backdoor access to the government [387].

2.4.3 Arguments against a Government Backdoor

In fact, what is indeed so special about the protection of phone data that merit a different treatment from data written in our paper phone book, letters we have at home, our call detail records, or our bank statements? Well, three arguments arise: the balance between privacy and security, the balance between security and security, and finally the limited effect of government backdoors.

Privacy versus security:
The first argument is a general one, and asks why we need to give up our privacy for the sake of security. Where is the balance between these two goals? Benjamin Franklin once said: 'Those who would give up essential Liberty, to purchase a little temporary Safety, deserve neither Liberty nor Safety.'[20] The argument, however, is not limited to encryption, but also to your 'offline' data.

Security versus security:
The second argument is more convincing, arguing that including a backdoor creates an inherent weakness in the data protection system. Even if you have full faith in the procedures that governments use to get access to the encrypted data, any malignant actor could try to exploit that weakness and find a way in through that backdoor. Given the continuous arm's race that is encryption and security, knowing that there is a weakness is likely going to attract hackers. The assumption that access to the backdoor is 100% secret is naive, thereby making the risk of abuse very large. A key under the doormat is not safe, neither is a government backdoor in our communication system [57]. This is also Tim Cook's argument, who was Apple's CEO at the time of the San Bernandino case. In a 2016 interview with ABC News he states the following [148, 476]: 'No-one, I don't believe, would want a master key built that would turn hundreds of millions of locks. Even if that key were in the possession of the person that you trust the most. That key could be stolen.' He adds: 'The only way to get information – at least currently, the only way we know –

---

[20] Although the quote perfectly fits the 21st-century discussion on liberty versus security in a digital age, the initial quote actually was not about pro-privacy, but rather pro-taxation and pro-defence spending in an 18th-century context [332].

would be to write a piece of software that we view as sort of the equivalent of cancer.' Apple deliberately does not retain the keys themselves.

So although a government backdoor would seemingly increase the security of the public, by allowing law enforcement to better track and capture criminals and terrorists, it simultaneously decreases the security of all other citizens, by exposing their data to blackmail or simply the loss of privacy. Not only personal privacy is at risk here, but also business data and secrets could be stolen, such as intellectual property or a company's strategy.

This argument is also followed by Jonathan Evans, the former director general of the UK's security agency MI5: 'While understandably there is a very acute concern about counter-terrorism, it is not the only threat that we face. The way in which cyberspace is being used by criminals and by governments is a potential threat to the UK's interests more widely' [181]. Looking further into the future, he points to the potential threat of cyberattacks against devices within an Internet of Things. The potential impact of not properly securing our vehicles, air transport, and even home devices is a great threat to society as a whole.

Apple is not the only one facing pressure from governments to allow for government access. Facebook is another large player that uses end-to-end encryption in its message services WhatsApp[21] (Facebook acquired WhatsApp in 2014). The data is encrypted on the sending phone and decrypted on the receiving phone. Neither Facebook nor the telco companies in between can see the plain text and only the encrypted data is sent. In a 2018 Facebook blogpost, the Global Public Policy Lead on Security argues a similar case: 'government officials who question why we continue to enable end-to-end encryption when we know it's being used by bad people to do bad things. That's a fair question. But there would be a clear trade-off without it: it would remove an important layer of security for the hundreds of millions of law-abiding people that rely on end-to-end encryption. In addition, changing our encryption practices would not stop bad actors from using end-to-end encryption since other, less responsible services are available. ... Cybersecurity experts have repeatedly proven that it's impossible to create any back door that couldn't be discovered — and exploited — by bad actors. It's why weakening any part of encryption weakens the whole security ecosystem.' [156]

---

[21] https://www.whatsapp.com/security/?lang=en

The futility of government backdoors in popular messaging apps and smartphones:

Even if Apple, Facebook, and all other major smartphone manufacturers and messaging apps would allow government access to the data, there will always be other ways in which bad actors will be able to communicate privately: be it through other apps that still work with end-to-end encryption without government backdoors, or simply by talking in person. Once such government backdoors are known (which would be the case when Apple would have built the software to remove the security feature), criminals and terrorist would very likely switch to other means of communication. Suppose even that no other apps could be found, a trained software programmer will be able to write the code for an encrypted messaging app rather easily. So once again a backdoor would weaken the security of the average law-abiding citizen, while not solving the problem of being able to access the data of malignant actors.

Former US Homeland Security Secretary Michael Chertoff reports: 'The bottom line is, if you look at both the terrorists in San Bernardino and the Boston Marathon bombers, they were family members. Most family members talk to each other face to face. The government doesn't have access to that after the fact' [252]. The futility of a backdoor is even subtly acknowledged in the proposed 'Brittney Mills Act' (which failed to pass in the Louisiana House committee in 2016) [456]. The proposal included a 2,500 US$ fine for the seller or leaser of a phone if the phone cannot be decrypted. But: 'There are exceptions to this rule in the case where a phone user may have downloaded a third party encryption app' [456].

2.4.4 And Now?

In the argumentation, a distinction is being made in data type, according to the source of the data. For non-digital personal data, such as letters you receive at home, printed bank records you might keep or printed invoices, there seems to be a consensus that law enforcement is allowed to have access to these and even to confiscate them with a proper court order. For digital personal records, a difference seems to be made according to whether the processing third party needs access to the content or not. Banks, for example, need to know the exact details of each payment: from whom and to whom they were sent, on what date, and so on. In those settings as well, there seems to be a consensus that law enforcement should be able to get access to the bank records of suspected terrorist or criminals (with proper court orders). Then why not for phone communications? Is it because such backdoors would be too easy to be accessed? The argument could also be posed in the other direction. Should telco operators encrypt all the text messages and phone calls? So that even with a court order, law enforcement would not be able to tap your phone records?

The issue of backdoors in messaging services and phones has really become a discussion point because of the Snowden revelations. In 2013, the at that time 29-year-old Edward Snowden revealed that many American hardware and software have 'backdoors' for American intelligence and law enforcement [34, 353]. As pointed out in a *New York Times* article, for a mobile phone company to survive in a global marketplace in this post-Snowden era, consumers need to be convinced that their data is secure [384].

Discussion 3

What do you think are the key drivers for making a distinction between the type of data and the possibility for the government to access these? Briefly discuss the arguments in favour of each of the following potential reasons:

1. The type of data: digital versus non-digital?
2. The fact that the third party already needs access to the data, and since the third party has the personal data, the government should also be able to access it?
3. The fact that there is much less transparency in how the personal data is obtained? A backdoor might be accessed in a manner that is too easy or without the scrutiny by the public or defending party?
4. Is it because for many people there is more personal information on their smartphone than lying around in their house?
5. Is it because of the revelations by Edward Snowden that without encryption there was a backdoor that intelligence services were able to access?
6. What are the limits when it comes to encryption and privacy? Should we all be able to keep all our personal data and messages secret, beyond the reach of law enforcement?

Encryption technology does not guarantee data protection. Apart from the continuous battle to keep encryption safe from attacks, there are still also issues such as meta-data and ways to circumvent encryption. Meta-data, such as the IP location from where a message is being sent, to and from whom messages are sent, and for how long calls are made [66], do not reveal the content of messages, but do provide personal information. Such data for example have reportedly been shared by WhatsApp to law enforcement in case of valid legal requests, such as (succesfully) helping in response to a kidnapping situation [323, 156]. Next to meta-data, encryption also does not help you when someone obtains access to your phone, while it is either not protected with an encryption key or when the key to access your phone is obtained. Once that happens, they see all your messages, even though they are sent encrypted. Nor does it help against backups or downloads of your chats that you might keep unencrypted. In an interview Tim Cook even points to this approach to access the personal data from the iPhone related to the San Bernardino shooting [148]: 'One of the things that we suggested was: take the phone to a network that it would be familiar with, which is generally the home. Plug it in, power it on, leave it overnight so it would back up, so you have a current backup ... backup to the iCloud.' This however was no longer possible in the San Bernandino case because the iCloud password was reset by investigators before, so the phone no longer backed up to the cloud.

Discussion 4

You're working at Messaging App NonExisting Inc., a popular app in your country.

1. You notice that your company is continuously sending all the messages between the customers of your company to the government. What would you do?
2. The child of the CEO has been kidnapped. The CEO asks you to quickly develop a backdoor so that law enforcement can access the last communication of his child. He promises that the software would be removed afterward. Would you? Why (not)?
3. Your own child or partner has gone missing. Would you develop the backdoor now?

Returning to the 2015 San Bernandino shooting. How did the FBI's demand for Apple to write software to unlock the iPhone 5C of one of the shooters end? After Apple declined to write the software, the case went to court. Finally, on 28 March the FBI announced it unlocked the phone and withdrew their request [53]. It was reported that they only found work-related information on the phone with no new revelations about the plot [424].

2.5 Bias

2.5.1 Bias: An Overloaded Term

The world is not perfect, and surely neither is data gathering. The data that a data scientist works on is seldom a perfect representation of the population on which the model is to be applied. The sample might have more data from easy to access groups, or historical data which no longer is representative due to a change in the environment or population. This becomes an ethical issue in two ways. Firstly, if it impacts the performance of the model on the sample versus the population. In that case, the performance evaluations that are provided on the test will differ from those that will be obtained when the model is deployed on the population, thereby yielding an incorrect picture of how well the model will perform. Secondly, if the sample is biased against certain sensitive groups, such as people of certain ethnic background, gender, religion, or age, the resulting model will likely include this bias as well. This could cause a negative discrimination against such sensitive groups.

Before continuing, a short discussion on the term 'bias': it is an overloaded term used for a variety of concepts in data science:

1. Bias in data sample: a non-representative sample of the population; we'll discuss this in the next subsection.
2. Bias of the data or the model against sensitive groups: this corresponds to the fairness issue, which will be treated throughout the book.
3. Bias/variance trade-off: the predictive performance of models is a trade-off between two errors [153]: the bias error caused by an assumption of the learning algorithm on the model, and a variance error which is caused from a not unlimited sample size, where the model is sensitive to small changes in the training set.
4. Bias in a linear model: the intercept is often also called the bias term. The reasoning is that if there is no data on the inputs (all zero), the result is the bias term.

All the uses of the term come back to the following general definition in the Oxford English Dictionary: an inclination in a certain direction.

---

bias: 'Inclination or prejudice for or against one person or group, especially in a way considered to be unfair.'

Oxford English Dictionary[22]

Whenever you talk about bias, it is important to make the context of this term clear. In data science ethics, this context is usually sample bias or bias against sensitive groups.

2.5.2 Sample Bias

Sampling is one of the inherent limitations of conducting data science. Due to a variety of reasons, gathering data on the complete population is often impossible. Data can only be obtained from persons who have provided consent; questionnaires and surveys cannot be obtained from the complete population; obtaining data can be expensive, and so on. When the sample is not representative, an error due to a non-random sample of a population is imposed.

In academic environments, students are regularly asked to fill out a survey or be part of an experiment, the assumption being that the students are representative of some of the population. However, only when this population is the set of students at that university is it likely not to impose a sample bias. Similarly, asking your own network to fill out a survey will introduce some sample bias as well, unless your network is a representative sample of the population you envision. Clearly defining your population and the sample will help in bringing forth the possible sample bias risks. A couple of examples follow next to demonstrate. Companies often face a similar issue, where the data at hand is limited to those of their (current and previous) customers. However, the population can be much broader, as the reject inference problem (see p. 74) illustrates.

So why is this bad? Well, because the following data science analyses can lead to wrong conclusions or models that impact certain groups in a negative manner. Let's consider a couple of examples.

---

[22] https://en.oxforddictionaries.com/definition/bias

Twitter is often used for a variety of analyses [255]. For a data scientist, Twitter data are a great, rich source of data from various types: network data in the form of who follows, retweets, or likes whom, textual data in the form of the tweets and short bio, sociodemographic data in the bio, and all of this with interesting temporal information so that behaviour and opinions of Twitter users can be observed over time. However, when making claims based on the modelling, we need to assure ourselves that the claim is made within the context of Twitter only. Because even if a representative sample of Tweets or Twitter users is chosen, there still remains a sample bias if the population corresponds to a broader audience of citizens or persons: social media users are not a representative sample of the population, just ask your (grand-)parents about their use. A 2011 study on a set of Twitter users representing 1% of the US population found that the Twitter users make a highly non-uniform sample of the US population [304]: Twitter users are (1) overrepresented in densely population regions, (2) predominantly male (although this bias seems to be declining over time), and (3) non-representative race/ethnicity distributions, for example oversampling of Caucasian users in various major cities and undersampling of Hispanic users in the southwest. A 2019 Pew Research Center study additionally found that Twitter users are younger, more likely to identify as a Democrat, highly educated, and having a higher income than US adults overall [415].

This bias can lead to misleading claims or interpretations of the analyses on such Twitter data. Trying to predict political preference, age, movie box performance or even stock prices using Twitter data should not be used to predict outcomes beyond this setting [161]. Predicting electoral outcomes based on Twitter data is one of such questionable analyses [161]. Figure 2.8 illustrates the point: suppose you want to predict the outcome of a US presidential race with two candidates: a democrat and a republican. A clever data science project is set up, and suppose you are able to get an accurate reflection of the political preference for all Twitter users as being a democrat ($Y = 0$) or a republican ($Y = 1$) voter. Then using these Twitter predictions to forecast election results will be misleading: even with the strong assumption of accurately being able to predict electoral preference for Twitter users, there is still the large bias of the Twitter population versus the total US potential voters population.

In World War II, a classified programme was set up to leverage the expertise of American statisticians to the war effort. A question that this Statistical Research Group (SRG) needed to solve was: how much to armour a plane [344, 126]. The more armour, the heavier and less manoeuvrable the plane; the less armour, the more vulnerable the plane becomes. Hence, this is

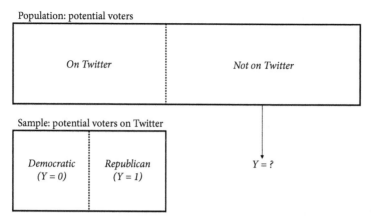

Fig. 2.8 Sample bias in Twitter users.

an interesting data science problem. The data provided came from American planes that returned from Europe and counted the bullet holes in the different sections. The fuselage was hit quite often (1.73 bullet holes per square metre); the engine not so much (1.11 bullet holes per square metre). Assuming the returning planes (the sample data) are a representative sample of all the planes, the answer would be to put more armour on the fuselage (Figure 2.9). Renowned mathematician Abraham Wald came to another solution: 'The armor, said Wald, doesn't go where the bullet holes are. It goes where the bullet holes aren't: on the engines.' [344, 126] If the planes were to be shot at all over, where are the holes over the engine casing? Wald's reasoning was that the missing holes were on the missing (shot down) planes. Thus, the parts with no bullet holes (the engines) were more important to protect: if those were to get shot, the plane would crash and not return. So additional armour was placed on the engines, where the bullet holes were found less frequently. This is another example of the implications of sample bias.

The third example of sample bias originates from the credit lending world, cf. Figure 2.10. Data science is used at major lending institutions to predict whether a customer will be able to repay the loan or not, and hence to decide on whether to grant credit or not. The resulting model would be used on all persons applying for a loan. So the population is everyone who applies for a loan at the bank, yet the sample is different: only data is available on those loan applicants who were actually given credit. The bank knows whether these people were able to repay the loan ($Y = 0$) or not ($Y = 1$). For those denied credit, the bank does not know the real outcome. These were probably not great customers, but some may have repaid the loan. The problem of figuring

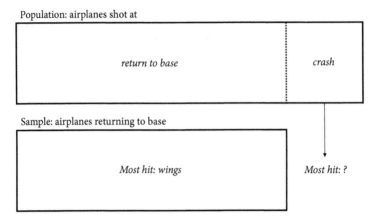

Fig. 2.9 Sample bias in planes returning to base.

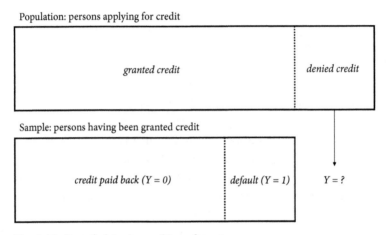

Fig. 2.10 Sample bias in credit applicants.

out what the outcome would have been for these persons that were denied credit is known as the 'reject inference' problem [100].

The most important aspects of sample bias are to be aware of it and to clearly limit the statements that are generated from the analyses and use of the data science models within those boundaries.

Not only can undersampling of sensitive groups be an issue, but over-sampling can be as well. Barocas and Selbst provide an example from the workplace [39]: if managers were to monitor employees of a certain sensitive group (for example, based on ethnicity or gender) in a disproportionate manner, mistakes made by employees would be logged in the company dataset at a higher rate for that group than for others. This practice could then become a self-fulfilling prophecy: the manager believes he or she is correct in focusing on that group as mistakes are reported at a higher rate, leading to even more focus on members of that group. By doing so, data science would be used in a wrong, unethical manner to confirm unwarranted prejudice.

So even when the under- or over-representation is completely unintentional, the resulting disparate impact warrants attention to this issue [39].

2.6 Cautionary Tales: Bumps, Gorillas, and Resumes

Sampling issues can bring about quite a commotion, and even large companies with brilliant engineers and an abundance of data and computing power still face these challenges. The two cases below illustrate the importance of transparency in the process, explaining models, and detecting and removing bias.

The previous examples show that sample bias can lead to the wrong conclusions. Consider the Street Bump app, introduced by the city of Boston to automatically detect and report potholes [98, 400]. Once the app is launched on a smartphone, the accelerometer can sense when a bump is hit, which is recorded together with the GPS location and sent to a city server [400]. As pointed out by Kate Crawford [98], such data gathering practices will obviously lead to less data from neighbourhoods with fewer smartphone owners, often consisting of people of elderly age and lower income [349]. It is reported that the city of Boston made efforts to address this data issue [98], yet it reveals that blindly working on gathered datasets can lead to a negative impact on sensitive groups in the population; in this case: less reported potholes and road improvements in poorer communities.

Similarly, a historic under-representation of certain racial or gender groups in your employee pool, because of a bias against these groups, can creep in the subsequent models. As we will discuss in Chapter 5, there are numerous cautionary tales on this subject, such as Amazon's recruitment tool that predicted which candidates would be suitable for an engineering position [102]. One way to do so is to set this up as a prediction task, using all candidates that sent their resume in the past, and using the words in the resumes as input features. All candidates that ever applied and were hired, get a positive label, all others a negative label. Amazon reportedly had built a similar recruiting prediction model, using resumes of a 10-year period. However, when reviewing the model, there tended to be a preference for male candidates [102]. Due to the historical bias towards male candidates, resumes that included words as *women's* (as in 'president of the women's chess club') or the names of colleges where only women can attend, were consistently downweighted by the model.

Amazon quickly abandoned the project when this bias came to light [102].[23] In face recognition, similar under-sampling of darker-skinned persons can lead to worse performance for images of this category. Buolamwini and Wilson found that two facial analysis benchmark datasets were overwhelmingly biased towards lighter-skinned subjects and that some commercial gender classification systems perform much worse on darker-skinned females [72]. Not accounting for this bias will lead to worse performance in various settings. Something even as simple as an automatic soap dispenser can go wrong: a 2017 video revealed how a soap dispenser only worked for lighter-skinned hands, not for someone with darker skin [260]. The technology was likely not properly tested on a representative (in skin colour) sample of persons. A similar problem has been reported on for some heart rate trackers [361], which of course can have more grim consequences.

Such a negative racial bias also exists elsewhere in the medical domain. Advances in predicting complex traits using genetic data have already led to the reported ability to predict breast cancer and type 1 diabetes risk better than current clinical models [289]. However, this capability is mostly limited to European descent patients only. The bias in predictive performance in favour of individuals of recent European descent clearly represents both an ethical and scientific issue, which the authors of a 2019 study in *Nature Genetics*, Alicia Martin et al., describe as 'the most critical limitation to genetics in precision medicine' [289].

Yet another example relates to the Google Photos app. This is an app where you can automatically group the pictures on your phone. For example, it will detect if you have a set of pictures from lakes, and will group them in a folder called Lakes, or a set of pictures with skyscrapers and it will group them in a folder named Skyscrapers. At some point it was revealed that the AI model wrongly put a picture of two black persons in a folder named Gorillas [7, 238]. Of course a big mistake by the prediction model. Google quickly reacted that it was 'appalled and genuinely sorry' [402, 41], and turned off the 'gorilla' prediction altogether to alleviate the issue. The chief social architect at Google apparently reached out to the one who posted the picture with the mistake, stating that 'different contrast processing needed for different skin tones and lighting ... We used to have a problem with people (of all races) being tagged as dogs, for similar reasons.' [238] and providing further explanations. Such a quick and transparent reaction is the right first thing to do.

---

[23] There is ethical merit in bringing such findings to light as these will caution other data scientists working on data-driven, automated recruitment on the risks of doing so.

Three years later, in 2018, *Wired* revealed that the Google system was still censoring the term 'gorilla' from searches and image tags. A spokesperson reportedly stated: 'Image labeling technology is still early and unfortunately it's nowhere near perfect.' [402]. And indeed, this can be a very difficult issue to deal with and shows the challenges when working with complex models using huge datasets. This illustrates that even large companies with abundant data science resources face these issues. The ability to explain (wrong) predictions can be of great assistance in such scenarios, see Section 4.4.4. That Google is taking these issues seriously is also reflected by the AI Principles that they published in 2018 [171], where a specific section is dedicated to the issue of fairness: 'Avoid creating or reinforcing unfair bias.' They further write: 'We recognize that distinguishing fair from unfair biases is not always simple, and differs across cultures and societies. We will seek to avoid unjust impacts on people, particularly those related to sensitive characteristics.' [171, 414]

2.7 Human Experimentation

Human experimentation is a longstanding scientific practice of conducting experiments on humans, to learn about humans. An experiment can be defined as 'an act whereby the investigator deliberately changes the internal or external environment in order to observe the effects of such a change' [388]. This is a common practice to gather data, and surely online. For example, by changing the message of an online ad to see the effect on clickthrough rates, or by changing the format of a website to see the effect on time spent on the website. Yet, the impact of experiments on humans can be much larger than simply clicking on an ad. Even though in medical research human experimentation naturally comes with ethical standards and thinking, in a non-medical data gathering stage (often online) such practices seem much less standardized.

We'll start with the historical background of common ethical guidelines for research on human subjects, to then compare traditional A/B testing with more problematic C/D testing, and finally a couple of cases from the online world where the experiments potentially had an effect on the mental health of the online participants.

2.7.1 Origin of Ethical Guidelines for Human Experimentation

Human experiments are what brought us many medical advances, including vaccines. The very first vaccine originates from the 18th century, when Dr

Edward Jenner observed that milkmaids with *cowpox* appeared to be immune to *smallpox* [50]. Dr Jenner decided to test his hypothesis by injecting some cowpox pus in the arms of a healthy eight-year-old boy and subsequently giving the boy smallpox. The story goes that indeed, the boy did not get sick from the smallpox [180]. The scientific paper he wrote on this, where he added a description of 13 other cases who contracted horsepox or cowpox and did not get sick after being exposed to smallpox, reportedly got rejected by the Royal Society, who suggested that he should cease his experiments [50].[24] He (luckily) ignored the advice and coined the cowpox material 'vaccine virus' (from the Latin for cow: vacca), thereby inventing the concept of vaccine. Although some ethical remarks can already be made here, much worse human experiments can be found in history.

The Tuskegee study started in 1932 in the town of Tuskagee, Alabama (US), and investigated the effect of untreated syphilis on black males [63]. A total of 400 syphilitic black men participated in the study, assuming they'd be receiving medical treatment. When penicillin became the widely agreed upon and available treatment for syphilis in 1951, still the men did not receive therapy, as to continue to observe the effects of untreated syphilis [430].[25] Only when the national press got wind of the study in 1972, did the experiment come to a halt. Somewhere between 28 and more than 100 of the subjects had already died at that time from syphilitic lesions. A panel later found that the study was 'ethically unjustified' as it failed to obtain informed consent from the subjects, and makes the case that penicillin should have been given to the men [63].

Some of the most horrific human experiments were conducted by the Nazis in World War II, with experiments that studied, among others, the human body's resistance to low pressure, malaria, mustard gas, and poison [461]. The so-called Angel of Death, Nazi doctor Josef Mengele, conducted heinous experiments on Gypsy children, twins, dwarfs, and people with abnormalities [95]. At the end of the experiments, the human subjects were killed and their bodies were analysed during autopsy. Of the 3,000 twins that Mengele experimented on, only 160 survived [258]. At the 1946 Nuremberg trials, 16 German physicians were convicted of crimes against humanity [95]. As a result of these revealed atrocities, the Nuremberg Code was established in the following year, with 10 points that describe the ethical rules for research involving human subjects [460]. These touch upon: informed consent without coercion,

---

[24] Yes, the study describing the very first vaccine was rejected by a journal; let it be a motivation for the academic reader who is facing a rejected submission.

[25] During World War II, several subjects received letters from the local draft board with an order to take treatment. At the request of the researchers, the draft board agreed to exclude the men from the required treatment [430].

the ability of the subject to withdraw from the experiment at any time and avoiding all unnecessary physical and mental suffering and injury [460]. The Nuremberg code was aimed at preventing a recurrence of such horrific experiment from ever happening again. But that didn't end unethical human experiments.

In 1966, Beecher published a seminal paper that identified 22 ethically questionable research projects [47], some of which were conducted at leading medical schools and published in well-respected journals [190]. The paper demonstrated that ethical experimentation with human subjects requires active thinking [405], and led to renewed groundwork for various ethical codes that we know today [190].

The 1964 declaration of Helsinki was developed by the medical community of the World Medical Association [470] and is an important document in the domain. The declaration has undergone various revisions, the seventh of which was of 2013. The basic principles talk about the respect of the individual, the rights of the individuals taking precedence over that of doing medical research to generate knowledge, and the need for qualified researchers [470]. Other principles include the minimization of risks and only conducting research on human subjects if the importance outweighs the risks to the subjects, the specific consideration of vulnerable groups, the need for research protocols and research ethics committees, and importantly, informed consent. The 1978 Belmont report is the basis of many ethical guidelines on human experimentation and focuses on three main ethical principles when conducting human research: respect for persons, beneficence, and justice [405].

There are clearly reoccurring recommendations and issues, which should guide us in our data gathering process. The first is that of 'informed consent', which is also part of GDPR. However difficult this can be, informed consent should be free and informed with objective information about the nature of the research, the potential consequences, risks, and alternatives, and should be obtained prior to the experiment [95]. In the Tuskegee study, the men were not aware of the potential treatment and devastation that untreated syphilis would have on their community. In Nazi Germany, consent (let alone informed consent) was totally absent. The second recommendation is to minimize the risk of the data subjects, and maximize the potential benefit. Finally, there is the need for oversight, not only at the beginning of a study, but ethical reflection throughout the experiment, especially if it concerns a study over multiple years. So go beyond the initial ethical approval of an ethics board: you might want to assign a person (or even a complete board) responsible to follow up and challenge the ethical implications during a data science project. Or you may want to add an ethical reflection in each report of the study.

2.7.2 A/B Testing and C/D Testing

In a typical experiment, we would want to discover what would happen if we change property X, and leave all other properties the same. To find this true causal relationship, one would need two parallel universes, which differ only in property X. What comes closest is a randomized A/B experiment.

A/B testing is a commonly used approach, where you perform experiments with two groups which differ in the treatment of certain attributes. For example, if you have two versions of a website for a conference, with two different colours, you can deploy both. Visitors are assigned to either version randomly, and some metric is defined (for example registrations) that measures success. This type of experiment is widely used in advertising as well, to determine which version of an online ad clicks or converts the best. But treating people differently can have serious consequences.

Of course, the emotional health of persons could be impacted, for example your happiness, without the user's knowledge of being involved in such an experiment. Such experiments can go beyond the simple change of the colour or layout of a banner or text, as the changes involve the use of data science models. Raquel Benbunan-Fich proposes the name C/D experimentation for such tests, where programming code is changed to manipulate results without forewarning, thereby intentionally deceiving the users [52]. In an experiment by OKCupid on 'love is blind' day, the pictures of the matches were removed for a while [319, 52], which Benbunan-Fich argues is a common A/B test [52]: the users are aware of the obvious change and as such there is no deception.

When should you be asking for informed consent? In a medical context, Austin Hill argues that when the patient will be subjected to discomfort or pain, informed consent is warranted [204]. In the digital setting, if there is a potential for negative impact, aim for an informed consent from the data subjects (and potentially model subjects as well). For example, if one changes the colours of an e-commerce site to see which version would lead to the most registrations, the impact is small, and a reasonable user would likely not expect to be asked for consent. If on the other hand, a dating site were to recommend persons who the site knows are likely bad matches, just to see what your reaction would be (see the next cases), the site is influencing the love life of the data subject, arguably requiring informed consent.

Assuming that the users provided *implicit* consent because of agreeing with long and complex terms of service agreements may be a valid legal argument, but it surely questions the validity of *informed* consent. A 2014 experiment in London showed that six people unwittingly agreed to give up their first born for eternity in exchange for free wifi, by agreeing with the terms and conditions when signing in, which included such a 'Herod clause' [147].

2.8 Cautionary Tales: Dating, Happiness, and Ads

2.8.1 OKCupid Testing of Their Match Prediction Model

OKCupid is an online dating site, where you can create your own profile, and a data science model can be used to predict who are the best matches for you. They reportedly have had this in place and wanted to test whether their prediction model actually worked well in practice [46, 68, 52]. So is the model indeed able to find good matches, or will people become a match, simply because of the 'power of suggestion', i.e. because OKCupid presents these persons as being a good match [46, 52]?

To test this, they created two groups [205]: group A had couples that had a low predicted match, and they were also told they were a bad match. But group B, also consisting of couples with a bad predicted match, were told they actually were a very good match. So they wanted to see if there would be a change regarding whether and how often the proposed matches talked to each other. And it turned out that indeed in group B, the persons were much more likely to start a conversation than in group A. At the end of the experiment, the users were notified about their actual predicted match scores [52].

There was quite some uproar when this was revealed [68, 468], even though the privacy policy did warn about potential research [205]. Arguably people felt that the dating site was playing with their happiness and love life. The at-that-time OKCupid president stated that this is simply how the Internet works: 'If you use the Internet, you're the subject of hundreds of experiments at any given time, on every site. That's how websites work.' [205, 68] And although this is true, there are ethical difference in testing which message of an ad clicks best or which colour for a website leads to the most registrations and conducting experiments that can have an impact on the emotional state of people. In a reaction by the *Washington Post*, the ethical implications of doing such experiments are summarized as [68]: 'If you're lying to your users in an attempt to improve your service, what's the line between A/B testing and fraud?'

The commonly agreed upon ethical practices concerning human experimentation from the medical domain don't seem to have become common in the digital world. Remember the need for informed consent, minimizing risk while maximizing the potential benefit, and ethical oversight, as the impact of digital experiments could be just as high as for medical experiments. So if your experiment has an impact on the emotional health of persons, then be sure to think through all the ethical implications of doing so. And at least make sure to have the explicit informed consent from users to be part of such experiments, however difficult that may be.

2.8.2 Facebook Contagion Study

Is your emotional state contagious outside of in-person interactions? This was the questions that a paper tried to address with a Facebook experiment on over 680,000 Facebook users, by reducing the emotional content in the news feeds [247]. Two A/B tests were conducted: in a first test group, people were shown less positive posts from friends, as these were (algorithmically) removed from the feed. In the control group no posts were removed in this way. It was observed that the group with less positive posts indeed posted more negative words in their following status posts as compared to the control group. The difference was reported to be significant, but relatively small [247]. In the second experiment, negative posts from friends were algorithmically removed, leading to more positive status updates. The study provided evidence that in such a setting your emotional state is indeed contagious, also without in-person interaction [247].

Users do consent to such experiments when they agree to its terms of service [208]. Yet, there was some uproar on the ethics related to this study as questions were raised about getting ethical approval for the study, and to what extent 'informed consent', as required in academic studies, was truly provided [208, 83]. This indicates once more that if you intend to gather data through a human experiment that might impact the mental situation of persons, you better think through all ethical implications, including obtaining informed consent, minimizing the potential harm to the data subjects, and ethical oversight.

2.9 Summary

All data science projects rely on data. The data gathering process needs to be fair to the data subjects and model subjects, in terms of privacy and discrimination against sensitive groups. Privacy tends to receive the most attention, where the European GDPR offers interesting inspiration on important definitions and principles. Personal data can be related to an identifiable person. The inverse—anonymous data—does not allow the data to be linked to an identifiable person, but such data are very hard to obtain. Pseudonymous data are what is often used, requiring additional data to link the data to a person. Encryption and hashing are key cryptographic techniques to deal with privacy: from creating pseudonymous or even anonymous data to more advanced schemes that we'll discuss in the next chapters. Governments often argue for backdoors in encryption standards, so they are able to retrieve personal data when appropriate (after a court order for example).

The transparency aspect of the data gathering process also needs to consider the privacy of the data subjects and the model subjects. This includes the aforementioned principle of informed consent: is the data subject and model subject informed about the data gathering and is consent provided? There should also be transparency in what data are gathered, for what purpose and for how long. Also the data scientist and manager require transparency, as to understand how the data are gathered. The data scientist needs to know how to ensure data quality and perform suitable data preprocessing and modelling, while the manager is the one signing off on the process, so he or she surely wants to know how this all occurs.

Let's briefly return to the initial story of this chapter, where Jenny was almost fired due to proposing the new data-driven business cases for online advertising, by (1) predicting product interest based on the obtained Facebook likes, (2) servicing music producers by turning on the microphone once in a while to listen to the music played, and (3) helping music event promoters by mapping the IP addresses to locations frequented. The GDPR principles point to the need of consent from the data subjects, which seems clearly missing here as the novel business cases were not thought of yet when the user downloaded the app. Purpose limitation is clearly violated as well: the initial purpose of using the data within the app is no longer the only purpose envisioned. Remember the La Liga fine for turning on the microphone once in a while, even though the obtained sound was hashed on the device before being sent to a server. So the investors were clearly right to be outraged, as there are major ethical transparency concerns with her proposals, and even legal ones if Jenny would have operated in Europe.

Bias is another important ethical concept, where sampling bias can lead to numerous wrong conclusions or an unfair treatment of sensitive groups. The cautionary tales indicate that this is a difficult issue to handle. Being transparent in the data gathering process, as well as in the response when something goes wrong, is an overarching theme in data gathering.

Finally, experiments are a common method to gather data on persons. History has provided numerous cases where this has gone horribly wrong, from Tyskegee to Nazi Germany. The common practices in medical experimentation can guide also typical experiments that data scientists conduct in an online or digital setting. The main ones are obtaining informed consent, minimizing risk for the data subject while maximizing potential benefits, and ensuring proper oversight.

Accountability requires that effective demonstrable procedures are in place, such as registering when informed consent was obtained and how. The cautionary tales indicate that the data scientists are not willingly unethical; rather, ethical reasoning has often not been a standard (or required) business practice.

3

Ethical Data Preprocessing

A newly appointed secretary of education wants to prioritize the digital agenda in higher education.[1] One of the projects she sets up is asking each university to provide a wide variety of data on their students to discover trends and needs in the student population. She specifically sends this request to the head of each university: 'We ask that all universities make the data on their students public, but to anonymize the names of the students by hashing them and not to include home address or other personal information in the dataset. For each student, we want the following fields to be included in the dataset: a hashed version of the student's name, the courses he or she enrolled in, his/her grades on these courses, days of absence in 2020 due to COVID-19, study program, nationality, date of birth, postal code and gender. In that way social science research can be moved forward, by finding patterns in this data, and universities could benefit from the discovered insights.' Spurred on by this request, the head of the university wants to leverage the students' data as well, by predicting who will end up in a *good* position after graduating. In this chapter we'll discover the different ethical pitfalls of this request: related to privacy, we'll see how removing identifiers is not enough to avoid being able to re-identify someone, the issue of working with sensitive information as grades or sick days, and how privacy can be enhanced by making the dataset k-anonymous, l-diverse, or t-close. On the additional subject of making predictions, fairness is a major concern, where one needs to consider what *good* means. The way in which this target variable is defined can already include bias in the dataset. We'll discover that simply removing sensitive attributes is also not enough to remove potential bias against sensitive groups, such as foreign students, and what methods exist to remove such bias from the dataset and avoid discrimination in the resulting predictive model.

---

[1] This is a fictitious story.

Data Science Ethics. David Martens, Oxford University Press.
© David Martens (2022). DOI: 10.1093/oso/9780192847263.003.0003

3.1 Defining and Measuring Privacy

When it comes to data, private or personal data boils down to data that can be linked to an individual. As one has gathered data, one needs to consider how to preprocess this data in order to store it in the proper manner (be it including personal, pseudonomized, or anonymized data). If a dataset is published, the risk of disclosing personal data needs to be minimized. Even if the data remain internal, the risk of data leaks and internal snooping by employees (or even court orders) requires this issue to be addressed. The ethical data scientist wants to be able to do useful analyses and modelling, while ensuring the privacy of the persons whose data are included. This can be done with Privacy-Preserving Data Publishing (PPDP) methods [155] such as suppressing instances or variables, grouping variables or values, and adding noise. These steps can be made on both a data instance or input variable level. Additionally, to ensure that the models resulting from the data don't show bias against sensitive groups, preprocessing analyses can be done that detect and remove bias that might be present in the data. Finally, in the target variable definition bias needs to be considered. Figure 3.1 provides an overview of these ethical data preprocessing steps, which are detailed next.

3.1.1 Suppressing, Grouping, and Perturbing

The most logical and common approach to make a dataset on persons more privacy friendly is by simply removing the personal identifiers, such as names, (email) addresses, and the like. This surely is a first step, but rarely solves the problem. As Samarati and Sweeney describe, next to explicit identifiers there

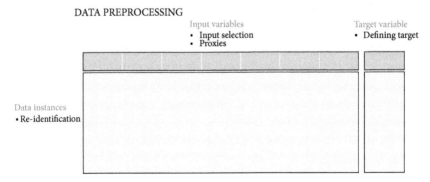

Fig. 3.1 Ethical issues related to data preprocessing.

are other distinctive features, which they term quasi-identifiers, which often combine uniquely and can hence be used to re-identify (or de-anonymize) persons [382]. Table 3.1 shows a fictional medical dataset with the full name as an explicit identifier, followed by three other variables (which are the quasi-identifiers) and finally the sensitive attribute: the medical diagnosis. One would clearly not want to reveal this dataset, as immediately the potential diseases of the included persons are revealed. By suppressing the name, we obtain the dataset depicted in Table 3.2.

But is the data now truly anonymous? If an adversary, let's call him Adrek, knows that Dirk is a 41-year-old man living in Antwerp (with postal code 2000), Adrek is able to link the first instance to Dirk, thereby revealing the diagnosis. So the combination of values for the quasi-identifiers allows us to link certain instances to individuals, thereby making the case that just suppressing the identifier is not sufficient. More methods are needed to further enhance privacy in the preprocessing step. This is what the US Federal Committee on Statistical Methodology calls 'statistical disclosure limitation procedures' [140], or methods to limit what personal data is disclosed. The purpose of such methods is to ensure that the risk of disclosing personal data becomes very small.

Table 3.1 Dataset with identifier (Name), quasi-identifiers (Age, Gender, Postal code) and Diagnosis.

Name	Age	Gender	Postal code	Diagnosis
Dirk Den	41	M	2000	Hernia
Eric Eel	46	M	2600	HIV
Fling Fan	22	F	1000	No illness
Geo Gen	28	F	1020	COVID-19
Han Hun	29	F	1000	HIV

Table 3.2 Dataset with suppressed identifier (Name).

Name	Age	Gender	Postal code	Diagnosis
*	41	M	2000	Hernia
*	46	M	2600	HIV
*	22	F	1000	No illness
*	28	F	1020	COVID-19
*	29	F	1000	HIV

In a well-known case, privacy researcher Latanya Sweeney purchased a voters list for 20 US$, which included name, address, postal code, birth date, and gender of each voter of Cambridge Massachusetts [421, 422]. She showed that 97% of the 54,805 included votes were uniquely identified based on only the full postal code and birth date [421]. By linking such public voter lists with published medical records, which also included the postal code, birth date, and gender of the included individuals, persons could easily be re-identified. One such case was William Weld, the former governor of Massachusetts who was re-identified and whose medical records were revealed. According to Sweeney, 87% of the US population could be uniquely identified using the quasi-identifiers postal code, birth date, and gender [422], thereby showing the danger of assuming that privacy is protected (or that the data is anonymous) by only removing explicit identifiers.

Further privacy-protecting methods are therefore needed. Grouping is a way to generalize the information of an individual, and can be done at an instance level, by aggregating instances into a cluster or at a variable level. The values of a continuous variable can be grouped into discrete values, while for nominal variables the values can be grouped into higher level concepts. Finally, perturbation adds noise in such a way that the statistics and patterns derived from the dataset will not differ much from when these would be derived from the original data (this comes very close to the definition of differential privacy which we'll consider in the next chapter, when dealing with privacy preserving data *modelling*). Adding noise can be done through injecting additive or multiplicative noise or data swapping (exchanging the sensitive attributes among instances) [155]. Although perturbation is a very simple and powerful method, it no longer keeps the data semantically correct, while grouping and suppressing do.

Returning to our example dataset, we can include grouping by making the age variable discrete, using equal-interval encoding with interval size of 10, and generalizing by mapping the postal code to the province (or state). This results in Table 3.3. In this dataset, Adrek can no longer uniquely identify Dirk, even if he knows that Dirk is a 41-year-old male from Antwerp. Adrek can assume though that Dirk is either instance 1 or 2, and hence is diagnosed with a hernia or HIV.

Now the question arises: how much of these privacy-preserving data publishing methods should you apply? At some point you'll have a completely randomized dataset or a dataset where all data instances have the same general value for all remaining variables. To answer this question, a definition on privacy of a dataset is needed: *k-anonymity*.

3.1.2 *k*-anonymity

Samarati and Sweeney proposed the concept of *k*-anonymity [382].

---

k-anonymity: 'A property of a dataset where for each combination of quasi-identifiers in the dataset, there are at least *k*-1 other instances with the same value combination.'

This implies that the information for each individual (data instance) cannot be distinguished from at least $k - 1$ other individuals in the dataset, with respect to the quasi-identifiers. In a *k*-anonymous dataset, the probability of linking a person to a specific data instance, through the quasi-identifiers, is at most $1/k$. Referring back to our example dataset of Table 3.3, we see that the dataset is 2-anonymous with respect to the quasi-identifiers: age, gender, and postal code: instance 1 and 2 have the same combination, while instances 3, 4, and 5 also have the same combination. These two groups are called *equivalence classes*: an equivalence class is defined as a set of records that have the same values for the quasi-identifiers. Adversary Adrek, who knows that Dirk is a 41-year-old man living in Antwerp, now only knows that there is a 50% probability that Dirk corresponds to instance 1 and thus has been diagnosed with a hernia, and a 50% probability that Dirk corresponds to instance 2 and has been diagnosed with HIV.

An important choice is how to define the set of quasi-identifiers. These are the variables that adversaries could potentially obtain through external sources. This is an important, yet difficult (and seemingly open) issue, where wrong decisions can lead to sensitive data being revealed, or an unnecessary information loss [155].

Table 3.3 Dataset with suppressed identifier (Name), generalized Age and Postal code.

Name	Age	Gender	Postal code	Diagnosis
*	[40–50]	M	Antwerp	Hernia
*	[40–50]	M	Antwerp	HIV
*	[20–30]	F	Brussels	No illness
*	[20–30]	F	Brussels	COVID-19
*	[20–30]	F	Brussels	HIV

Transforming a dataset to a k-anonymous dataset can be automated, and a wide range of algorithms have been proposed to do so through suppression and grouping, such as Sweeney's Datafly system [420]. For the interested reader, Fung et al. provide an interesting overview of such methods [155]. The goal of these anonymization methods is: given a dataset, obtain a k-anonymous version of that dataset with minimal information loss as fast as possible. Definitely for large datasets, greedy algorithms need to be employed, as it has been shown that finding the optimal k-anonymous version is an NP-hard problem [298].

This definition formalizes the issue of privacy in the preprocessing step nicely, as it guarantees that each instance in a k-anonymous dataset cannot be distinguishable from at least $k - 1$ other instances, even if the instances are linked to external information. However, two attacks are still possible, where the value for the sensitive attribute can be identified.

3.1.3 Homegenity and Linkage Attacks

Protecting the identity is different from protecting the sensitive attribute values. Even though an attacker will not know whether Dirk corresponds to instance 1 or 2, the sensitive value is revealed if both instances have the same value for the sensitive variable. Such a homogeneity attack [274] is illustrated in Table 3.4, where Adrek will infer with 100% certainty that Dirk has been diagnosed with HIV, even though he doesn't know if Dirk corresponds to row 1 or 2.

A second issue with k-anonymity is when an adversary has access to two datasets, with the sensitive variable present in both. This scenario would make the data vulnerable for a so-called linkage attack [155, 274]. Let's consider Table 3.5, which is inspired by the examples of Fung et al. [155]: suppose the

Table 3.4 Homogeneity attack on 2-anonymous dataset.

Name	Age	Gender	Postal code	Diagnosis
*	[40–50]	M	Antwerp	HIV
*	[40–50]	M	Antwerp	HIV
*	[20–30]	F	Brussels	No illness
*	[20–30]	F	Brussels	COVID-19
*	[20–30]	F	Brussels	HIV

Table 3.5 Linkage attack with additional dataset: if Adversary Adrek knows Fling is a 25-year-old woman from Brussels and in both datasets, Adrek can infer that Fling has HIV.

Name	Age	Gender	Postal code	Diagnosis
*	[40–50]	M	Antwerp	Hernia
*	[40–50]	M	Antwerp	HIV
*	*[20–30]*	*F*	*Brussels*	*No illness*
*	*[20–30]*	*F*	*Brussels*	*COVID-19*
*	*[20–30]*	*F*	*Brussels*	*HIV*
Name	**Age**	**Gender**	**Postal code**	**Diagnosis**
*	[40–50]	M	Antwerp	Hernia
*	[40–50]	M	Antwerp	HIV
*	*[20–30]*	*F*	*Brussels*	*HIV*
*	*[20–30]*	*F*	*Brussels*	*Hernia*
*	*[20–30]*	*F*	*Brussels*	*Heart attack*

data of two hospitals is revealed, each 2-anonymous and both with the same quasi-identifiers and sensitive attribute. If Adrek knows that Fling is a woman in her 20s from Brussels and has been a patient in both hospitals, Adrek can infer from these datasets that Fling has been diagnosed with HIV, because if we look at the diagnosis of the last three rows in each dataset, we observe that only the HIV diagnosis is present in both, so that must be the record corresponding to Fling. This attack requires more extended background knowledge, knowing that Fling is in both datasets, and that the sensitive attribute is present in both. The way in which Sweeney was able to reveal the medical records of the former governor is another example of such a linkage attack [422].

3.1.4 *l*-diversity and *t*-closeness

If we consider the homogeneity attack, one solution is to limit the size of the dataset, so that Adrek doesn't know whether Dirk is actually in the dataset or not. But suppressing instances might lead to unwanted reduction in the informativeness of the dataset. An extension of *k*-anonymity that aims to deal with this issue is *l*-diversity, proposed by Ashwin Machanavajjhala and his co-authors in 2007, by promoting the diversity of sensitive values within each group of *k* (or more) indistinguishable (when it comes to the quasi-identifiers) data instances [274].

> *l*-**diversity**: 'A property of a dataset where for each equivalence class in the dataset, there are at least *l* well-represented values for the sensitive attribute.'

Different definitions exist on what well-represented entails, the simplest being unique: so there should be at least *l* distinct values for the sensitive attribute for each group of instances which have the same values for the quasi-identifiers. If we look again at Table 3.3, we observe a dataset with 2-anonymity and 2-diversity. The dataset of Table 3.4 has less anonymity: 2-anonymity and only 1-diversity, and would require more grouping and suppressing of values and variables in order to make it 2-diverse.

Unfortunately, privacy concerns remain after ensuring *l*-diversity [268, 155]. Consider the scenario that one value of the sensitive attribute is much less common than the other, for example a positive COVID-19 virus test versus a negative one. Suppose you have an equivalence class with two data instances, one being positive and one being negative for COVID-19. If in the overall population only 1% has positive COVID-19 cases, then knowing that Dirk is in this equivalence class changes our belief that Dirk has COVID-19 from 1% to 50%. This leads to yet another privacy definition, proposed by Li et al, called *t*-closeness [268], which requires that the distribution of the sensitive attribute in each equivalence class is close to the distribution of the sensitive attribute in the complete dataset, with closeness defined by some distribution distance metric and a threshold *t*. Notice that enforcing this will make the data even more general, once more illustrating the balancing act between utility of the data and privacy of the data subjects.

3.2 Cautionary Tales: Re-identification

Behavioural data provides evidence of actions [396] that we take in a digital world. Think of visiting locations, making payments, making search queries, liking Facebook pages, or visiting webpages. For such data, each unique potential action, be it a location, an account number, a query, a Facebook page, or a webpage, is represented by an input variable. This variable is typically binary: one if the action is taken, and zero otherwise. One can quickly imagine that a few actions can identify an individual. For example, who besides me visits my son's daycare in Antwerp in the morning, the University of Antwerp during the day and a house in Berchem in the evening? Or similarly, who besides me makes payments to both the specific daycare in Antwerp, a local coffee shop near the university, and a small bookstore in Berchem? This is both the potential and danger of such fine-grained, very high-dimensional data: because it is so revealing of one's interests, it can be used to build accurate prediction models for a range of applications, from mobile advertising [365] and credit scoring [432] to predicting personality traits [246], but can also be used to re-identify persons, as the next cases demonstrate.

Grouping could entail the aggregation of all locations in the Antwerp area to one super-location 'Antwerp', or all unique account numbers from bookstores to a single 'bookstore' account number. Suppression could be implemented by removing all unique actions that occur less than a certain frequency. But both dimensionality reduction methods will very likely lead to a (substantially) reduced predictive performance, when used in predictive models [227]. Aggarwal specifically studied the effects of dimensionality on k-anonymity methods [4]. He finds that when a dataset has a large number of dimensions which can be regarded as quasi-identifiers (which is the case with behavioural data), one needs to choose between suppressing most of the data and losing the wanted level of anonymity. The following cases provide cautionary tales to illustrate the ease of re-identification when publishing behavioural data.

3.2.1 Re-identification based on movie ratings

Netflix is a subscription-based streaming service for movies and TV series that initially started off as a DVD rental (by mail) business. In 2006, Netflix announced the start of the 'Netflix prize': whoever was able to improve the accuracy of Netflix' recommendation model in predicting ratings (measured by the Root Means Squared Error) by 10%, would be awarded 1 million US$ [9]. A dataset was made public with over 100 million ratings (1 to 5) from about 500,000 Netflix users who rated movies between December 1999 and December 2005. Each data instance consisted of a user id, movie id, date of grade, and grade. The competition started in October 2006, and ran until July 2009, when a team finally achieved the required performance improvement to win. In those nearly three years, over 5,100 teams from over 185 countries participated in the contest [321]. The submission that won the prize consisted of three teams that joined their efforts, named 'BellKor's Pragmatic Chaos'[2]. In

---

[2] The second team, 'The Ensemble', reportedly obtained the same accuracy improvement but submitted their results 20 minutes after the winning team, thereby narrowly loosing the 1 million US$ prize [320].

2010 Netflix announced that it would start a second Netflix prize, but this was later cancelled due to privacy concerns and uproar.

Although the dataset was said to be anonymous, where only one tenth of the complete dataset was included, and the data was 'subject to perturbation' [315], major privacy concerns emerged. Two researchers at the University of Texas, Narayanan and Shmatikov, showed that by linking the supposedly anonymous dataset to auxiliary external data, users could be identified [314, 315]. In this linkage attack, they specifically looked at the Internet Movie Database (IMDb) where users can comment and rate movies [314, 315]. When someone rates a movie on both Netflix and IMDb, the authors assume that the date one rates a movie on Netflix is strongly correlated with the date one rates the same movie on IMDb. So when I watched and rated the movie *Top Gun* on Netflix, and rated the same movie on IMDb, this probably would have been around the same date. These dates don't need to be exactly the same to link the data, but rather one needs to know around what date a movie was watched and rated. This principle is illustrated in Table 3.6, where a (fictitious) anonymous Netflix user can be linked to the (fictitious) IMDb user johndoe90 as they both rated the same movies on the same day, give or take a day. By linking the data instances, we can now observe that johndoe90, described as being from Antwerp and a Keanu Reeves fan in his bio, also watched *Fahrenheit 9/11*, *Jesus Of Nazareth* and *The Gospel of John*. It's immediately clear that enforcing some form of *k*-anonymity is quite difficult, given the dates of the ratings. And indeed, Narayanan and Shmatikov find that even 2-anonymity would destroy most of the information contained in the dataset [314]. They

Table 3.6 Linkage attack with an additional dataset: by matching the movies rated on both IMDb and Netflix on about the same date, it can be revealed that user *johndoe90* also watched *Fahrenheit 9/11* on Netflix, potentially revealing his political preference, as well as *Jesus of Nazareth* and *The Gospel of John*, potentially revealing his religious preference.

Netflix		IMDb		
ID	Movies rated	ID	Movies rated	Bio
*	*A.I.* on 6/6/04 *Bullhead* on 1/1/03 *The Pledge* on 8/1/01 *Fahrenheit 9/11* on 3/15/04 *Jesus of Nazareth* on 10/21/00 *The Gospel of John* on 5/22/04	johndoe90	*A.I.* on 6/7/04 *Bullhead* on 1/2/03 *The Pledge* on 8/1/01	From Antwerp, Keanu Reeves fan

further describe that when a 3-day error is allowed, 96% of the Netflix users in the dataset can still be uniquely identified. For 64%, two ratings and date of rating are sufficient for complete de-anonymization. So for the users who watched Netflix, and rated movies on IMDb, the available personal information on IMDb became linked with the complete Netflix viewing history prior to 2005. The personal information includes the user name, but can also include the person's website address or personal biography. Yet another example that shows that simply removing personal identifiers does not necessarily lead to anonymous datasets.

So what if you're able to identify a person in the Netflix dataset? It turns out that your movie-watching history might reveal political, sexual, and religious preferences. Something that might not be possible with only the IMDb dataset (for example because you don't reveal on IMDb all the movies you have watched). Narayanan and Shmatikov argue that if one were to know the opinion of someone about the movies *Power and Terror: Noam Chomsky in Our Times* and *Fahrenheit 9/11*, the political preference could be inferred [315]. Similarly would the ratings of *Jesus of Nazareth* and *The Gospel of John* reveal the religious preference, and positive ratings for movies with predominantly gay themes, such as *Bent* and *Queer as Folk* could be predictive for a user's sexual preference. Kosinski et al similarly have shown that what you like on Facebook can be used to infer your political or sexual preference or even alcohol use [246].

Even if you are thinking now that you, or an average Netflix user, would not mind that their historical movie preferences are revealed, Narayanan and Shmatikov rightly state that the privacy question is not whether the *average* Netflix user would care; rather the question is whether *any* of the Netflix users would care if their complete movie history is known [314]? And of course some did object. A lawsuit was filed against Netflix by Jane Doe [232]. Jane is reported to be a lesbian Netflix user, whose sexual preference is not a matter of public knowledge, 'including at her children's school' [232, 232]. She watched movies in the Netflix category 'Gay and Lesbian', which could reveal her sexual preference. In the lawsuit, the plaintiffs claim that if her sexual preference were to be publicly known 'it would negatively affect her ability to pursue her livelihood and support her family and would hinder her and her children's ability to live peaceful lives within Plaintiff Doe's community' [232]. Furthermore, the plaintiffs argue that Jane Doe 'will be irreparably harmed by Netflix's disclosure of her information in its upcoming contest'. Ultimately the lawsuit was settled and the second Netflix prize was cancelled [271, 338].

The ability to re-identify someone based on the pre-2005 Netflix movie watching history also holds implications for future privacy [314]: any user whose identity is revealed through their movie ratings cannot disclose any non-trivial information about his or her movie preferences in the future, because that could be used to link the record to his or her identity. So anonymously commenting somewhere that you liked or disliked a certain (pre-2005) movie can lead to your identity being discovered. And the Netflix prize data is not going away: as it has been made public, many copies have been made and it is likely to be publicly available for a long time.

A solution might be not to reveal the movie names in the dataset. But any recommendation algorithm that used content information about the movie would no longer be usable. Another solution would be to include more perturbation and grouping, but this comes once more at the expense of (substantially) reducing the utility of the resulting dataset and resulting data science analyses. Only allowing selected researchers to work on the data, after signing a Non Disclosure Agreement (NDA), would allow for a competition without the need to make the dataset publicly accessible, though likely not all 5,000+ teams would have participated in the Netflix prize then, not to speak of the administrative burden that this would bring about. Furthermore, if only one participant leaked the dataset (or it was hacked), the same privacy issues would emerge. This once more comes back to the continuous balancing act between privacy and utility.

Finally, note that the Netflix prize has spurred a lot of research in the area of recommender systems, and led to a large, real-life dataset for researchers to experiment on. The idea of organizing a data science contest was also forward-looking; think for example of the popular online Kaggle community that was started in 2010, where datasets are published for data scientists to work on. One of the winners of the Netflix prize, Volisnky, accentuated the positive intentions of Netflix: 'I think it was really unfair because Netflix behaved really well and were good stewards of their customers data' [223]. This demonstrates that even with the best intentions, it is important to be aware of the ethical risks of data science.

3.2.2 Re-identification Based on Search Queries

Another well-intended cautionary tale comes from AOL in 2006 [338, 331, 188], an Internet giant at the time. In an effort to foster academic research, they published 20 million search queries originating from over 650,000 users (reportedly about one-third of all searches conducted through the AOL network

Table 3.7 Example entries from the AOL search queries logfile: the entries are fictional; the structure is as in the released dataset.

ID	Query	Query Time	ItemRank	ClickURL
1365	baby names	2006-05-05 09:16:10	2	www.babynames.com
1365	daycare berchem	2006-05-06 10:11:55		
1365	data science ethics	2006-05-06 22:10:06	1	www.dsethics.com
4569	passover cakes	2006-04-14 14:16:22	3	www.recipezaar.com
...				

in that time period [331]) over a three-month period. Once more, personal data such as usernames and IP addresses were suppressed. Each user was assigned a random id number, so all searches of the same user could still be identified. Next to the search query, also the date and time of the search query were included, as well as the address of the online webpage that the user clicked on [188]. Table 3.7 provides some fictitious data instances for the provided data structure.

The goal of publishing this dataset was a well-intended attempt to reach out to the academic community, and to 'embrac[e] the vision of an open research community' [338]. The data was to be used by search engine researchers in academia, for a non-commercial use only [478].

However, very quickly there was a lot of public uproar over the privacy implications of releasing this data. Firstly, the fine-grained and revealing nature of the data led to the ability to re-identify some users. Secondly, just as for the Netflix re-identification case, one could question whether that is really so bad? Well, it is, given the very sensitive, and sometimes disturbing, nature of what we search for online.

Let's start with the re-identification. Reporters Barbaro and Zeller from the *New York Times* were able to (re-)identify user 417729 as Thelma Arnold, based on her search history [35]. Arnold allowed her identity to be revealed by the reporters and confirmed she was the author of the searches. Some queries that allowed her to be identified included: 'landscapers in Lilburn, Ga', 'homes sold in shadow lake subdivision gwinnett county georgia' and several people with the last name Arnolds [35, 338]. Some other, slightly awkward queries by her included '60 single men', 'dog that urinates on everything'; but also warm-hearted queries appeared as 'school supplies for Iraq children'. She apparently was shocked to hear about the data publication: 'My goodness, it's my whole personal life' she said [35]. One does not need a lot of imagination to figure out that many users could be re-identified in this way.

The queries also revealed some very disturbing and sensitive thoughts, such as the query 'how to kill you wife' (including the typo). User 3505202 queried about 'depression and medical leave', another user searched for 'fear that spouse contemplating cheating' [35]. Yet another user seemed to work in finance, was searching online about high blood pressure, and looked up escort services in several cities he was presumably visiting [188]. It was further revealed that there were thousands of sexually oriented queries, including some about child pornography.

Upon these revelations, AOL swiftly removed the dataset from its website. A spokesperson said that the publication of the data was a violation of internal policies and issued a strong apology: 'It was a mistake, and we apologize.' [188] AOL reconsidered the length of time it holds the search queries that customers make and re-educated its employees about the sensitivity of the data [478]. The researcher that released the data as well as his supervisor were fired and the CTO resigned [478]. Even though AOL quickly removed the dataset from its website (three days after its release), copies continued to circulate online.

Some potential solutions include once more further suppression and grouping, but again at the expense of the utility of the data. Further limiting the time of storage of data (as AOL indicated themselves), and limiting and logging access to the data are other measures that could be helpful as well. To still be helpful for the academic community, the dataset could have been shared only with a limited number of (trusted) researchers, after signing an NDA.

Some academics (and even non-academics) might wonder: if you obtain access to such data, should you use it? Once more a balancing act has to be achieved between the ability to move research forward by developing and validating algorithms using real-life data versus the extent to which personal and sensitive data are contained in such data and the manner in which they have been obtained. In the digital age we live in, most massive datasets originate from such large (technology) companies. Researchers covet such datasets, as it provides them with a unique asset in their research. How else to demonstrate that your novel idea for an algorithm to better search the Internet, make predictions, or recommend movies actually works in real life? Yet, using such data can taint your own research later on as well. Related to the AOL case, Professor Jon Kleinberg provides the following guidance in a related *New York Times* article: 'The number of things it reveals about individual people seems much too much. In general, you don't want to do research on tainted data.' [186]

Google provides the 'My Google Activity' tool, where you can 'rediscover the things you've searched for, read, and watched', and delete your activity if wanted. As stated by Google, only you can see this data. Have a look at your own search history, through https://myactivity.google.com. Now consider whether someone would be able to identify you, whether you'd be embarrassed, or worse, whether this might have a negative impact on your family or professional life, if your searches became public.

An interesting related side story is that of the 2006 subpoena by the US Justice department, demanding four large search companies to hand over query data of millions of its users in an effort to uphold an online child pornography law [478]. Google declined to turn over the query data, one reason being it could expose identifying information about its users, and ultimately won the right to withhold the query data [187]. It was only required to turn over a portion of anonymous search results (the resulting website addresses from search queries) but not the queries itself [478, 374].

3.2.3 Re-identification Based on Locations

Location data is a rich data source that has been used for research and commercial (mainly app) purposes. This data might well be the most sensitive of all behavioural data types, as the revelation of the locations you visit can easily expose your identity. The location you are most frequently at during the night is likely your home address, while the location you visit most often during office hours is likely your work address [357]. Other locations you visit can reveal much personal and sensitive information as well. Such data can easily be retrieved from our mobile phones: either through the location of connected cell phone towers that is available at telco operators, GPS (latitude/longitude) coordinates when we allow location data to be sent by our phone apps, when we explicitly check in on apps like Foursquare or Facebook, or through the logging of visited IP addresses of WiFi networks. Apps on your mobile phone often use location data, for example weather apps that want to update their predictions based on where you are at that moment, apps that provide driving directions or sports apps that track your running or biking routes. But location data is often also used for other purposes, such as targeted advertising, and sometimes even sold. New York Times reporters were able to look at a 2017 database with the sampled whereabouts of more than a million phones in the New York area [445]. They report that at least 75 companies receive 'anonymous' precise location data from such apps, whose users enable location services [445]. And the same cautionary tale unfolds: several identities could be revealed, as well as startling and disturbing information on users.

The New York Times reporters were able to identify Lisa Magrin, a 46-year-old math teacher. (She allowed the New York Times to reveal her identity.) Lisa could be identified as she was the only person who commuted daily from her house in upstate New York to the middle school where she works. Not only her home and work location were revealed, the dataset showed she visited a Weight Watchers meeting, she stayed at her former boyfriend's home, visited the gym, and even a doctor's appointment was included. 'It's the thought of people finding out those intimate details that you don't want people to know.' [445] Imagine being able to identify all persons (or better mobile devices) that visit locations such as a military base, a psychiatrist office, an AA meeting, a nuclear power plant, schools, churches, mosques, and so on. Finding the home address of each device could be done by looking what the most frequently occurring location is of that device at night [357]. Simply knowing their identities would already be disturbing, but this knowledge could lead to unethical and even illegal practices such as blackmail.

Not only can apps track your location, but telco operators can do so as well, based on triangulation of connected carrier's antennas. In a study by Yves-Alexandre de Montjoye and his colleagues, human mobility data of 1.5 million individuals over a time period of 15 months was analysed [106]. The results indicate that four spatio-temporal points (knowing the location and time of visit up to the hour) were enough to uniquely identify 95% of the included individuals. Furthermore, aggregating the data to a more coarse-grained level of location (by aggregating the reception area of two antennas) and time (using a window of increasing number of hours) did not provide much more anonymity: for example when allowing a resolution of 5 hours and aggregating the region of 5 antennas into one location cluster, still more than 50% of all users were uniquely identified with four randomly selected locations [106]. Aiming for 2-anonymity hence would require much aggregation, thereby losing the granularity of the dataset which makes it valuable for data science.

And even then, as we saw earlier, k-anonymity is still open to privacy attacks: consider a telco operator that wants to leverage its dataset by providing market reports to businesses. A casino might ask how many visitors come from neighbouring cities or countries, one answer being: 50 persons from zip code 2222 visit the casino in the weekend. If you know an individual who lives in zip code 2222, and know that this zip code only had about 100 residents, you'd be able to infer with 50% certainty that this person frequents a casino. If the zip code would only have 50 residents, you would know for sure they all visit the casino (homogeneity attack). What if the telco operator reported how many citizens from each zip code joined a protest, or how many visited an abortion clinic? Once more, attention should be paid to two aspects: making sure no one can be identified, and secondly avoiding the revelation of any potentially disturbing or sensitive information (such as visiting doctors' offices or casinos).

Location data can also be obtained without the use of mobile devices. New York City released the data of over 170 million individual taxi trips, after a request was filed by an open data activist under the US Freedom of Information Law [466]. Inadvertently the home addresses of the taxi drivers were revealed, as well as personal information about individual customers, as the pick-up and drop-off locations and times of each taxi trip were included in the dataset [200]. If one observes that a taxi picked up a person at a certain home address, the drop-off location can easily be found for the person living at that home address. As one commenter on Hacker News stated: 'can you imagine someone just plotting all the trips from a single gay bar? Listing off all the connected residential addresses? And not only that, any subsequent trips home from those addresses the next morning?' [200, 1] Even worse, it turned out that the hashing of licence plate and taxi medallion numbers was easily reverse-engineered [200]: instead of mapping each licence number to a random number, it was hashed with the common MD5 hash. Knowing that licence plate numbers are all six- or seven-digit numbers (all starting with the number five), a rainbow table could be made for all possible licence plate numbers. By looking up the hash in this table, the licence plate numbers (and similarly the taxi medallion numbers) could be revealed to the public. Mapping medallion numbers to driver names is reportedly easily done by searching online [200].

Let these cautionary tales warn you that simply removing device ID, IP address, or name does not necessarily make your data anonymous, and certainly not when dealing with behavioural data. Such data can lead to the revelation of identities and disturbing facts. The concept of k-anonymity takes away from the utility of such fine-grained behavioural data to make predictions, recommendations or to simply find interesting patterns. Therefore one should avoid publicly revealing behavioural data, assuming it is anonymous by removing personal identifiers such as name and address. There are ample of good reasons to share such data with researchers: to foster research, for transparency, to advance society; yet do so with trusted research groups only, and after signing agreed upon data protection and non-disclosure agreements. If general statistics about such data are to be revealed (cf. the example of a telco operator providing business reports based on location data), the use of differential privacy is recommended. Even the simple storing of behavioural data at a company or at the government requires extra careful thinking, as the data can be hacked, subpoenaed, or employees might snoop around. In those cases, remember the GDPR article 5 principles such as data minimization, purpose limitation and not keeping the data longer than needed.

Discussion 5

Suppose that you have gathered a dataset of Facebook like data of 10,000 Facebook users. This dataset was gathered in an ethical and legal manner, among others with informed consent.

1. A research group at a renowned university suggests you to make the data public, but to anonymize the names of the users by hashing them. In that way data science research could be helped by testing new algorithms and metrics.
   (a) Would you?
   (b) What would be potential pitfalls or public outcries? Think of re-identification or finding disturbing patterns.
   (c) Would there be a way where this dataset could be leveraged for data science research, while ensuring ethics?
2. A journalist asks for access to the dataset, promising not to make the content possible. The journalist wants to see what the dataset would reveal. Would you? Why (not)? Consider the same questions as in (1).

3.3 Defining and Selecting Variables

3.3.1 Input Selection

Input selection encompasses all previously seen issues and methods, trying to ensure that privacy is respected, and minimize discrimination against sensitive groups. Data minimization is also a concept that should be taken into account here. As Sweeney has demonstrated, the combination of a few variables, such as date of birth, gender, and zip code, can allow persons to be uniquely identified [422]. Having other fine-grained variables, such as products bought, job, or street name, will lead to included persons being re-identified more easily. As before, there is a trade-off between privacy and utility. So there might be good reasons to still have many variables, so as to have well-performing models. Yet, the relevance and predictability of each variable should be motivated, among others, using input selection procedures. Earlier seen methods such as hashing, encryption, and k-anonymity can also be of help to add more privacy.

Next to privacy, fairness requires us also to look into potential discrimination against sensitive groups. If we don't want to treat persons differently based on certain sensitive attributes, these variables should not be included as one of the input variables. Yet, in order to measure and remove bias against such groups, the attributes will be needed. Also beware of proxies of such variables. Creative thinking might reveal good reasons to include certain variables, such as location data or job type, but beware of the danger of discrimination this might bring about, for example against race and gender.

3.3.2 Defining Target Variable

Defining the target variable is inherently subjective. It therefore requires careful thinking about both the business and ethical implications. In predictive modelling, this variable is what the model will try to predict, so whatever bias is in the definition or initial measurement of this variable, risks to be included in your final model as well. As Barocas and Selbst stated, 'a target variable must reflect judgements about what really is the problem at issue.' [39] They specifically discuss this issue in terms of hiring decisions, and motivating this can stem from administrative cost reduction, improving sales, or innovation. And indeed, each of these will likely lead to a different definition for the target variable. So let's assume we want to build a data mining model to predict a 'succesful' hire. But how to define this? Is this anyone who stayed with the company for at least five years? Employees that advanced within the company to a managerial level? Is it just anyone who got hired in the past? All of these can have ethical repercussions: if the company had a bias against women for such positions, the model will include this bias as well. If female employees took maternity leave during their first years at the company, is this accounted for in the definition of 'successful' hire? This is where ethical thinking about the specific definition is needed, by attempting to discover scenarios where certain groups might be unintentionally discriminated against.

So target variables can be biased in its definition or measurement, but they can also concern an attribute which is too intimate and personal, leading to what Crawford and Schultz call 'predictive privacy harms' [99]. Consider trying to predict pregnancy. This might be warranted in a medical context, but is it in case of targeted advertising? Similarly, Kosinski et al. have shown that personal political, sexual, and religious preferences can quite accurately be predicted from Facebook likes [246]. But should you? This likely will depend on the context and informed consent of the user. Similarly, Sandra Matz and her co-authors revealed how targeted advertising based on predicted psychological traits can lead to more clicks and purchases compared to unpersonalized targeting [293]. In their study, they discuss the potential for doing good, for example by targeting highly neurotic persons who display early indications of depression, with health advice or information; but also the potential unethical use, by using the prediction to keep citizens from voting, as one example [293].

As a final example, imagine trying to predict who will get engaged, based on Facebook likes, which might be very interesting information for wedding planners and jewellery stores. Suddenly getting ads on your screen for diamond rings might feel creepy, but imagine having your partner standing next to you when the ad appears, or that your partner actually also gets to see the ad. Likely nothing illegal is happening in this case, yet the ethics of such practices are more dubious.[3] Once more, the context determines the ethical data science practice related to defining the target variable.

3.4 Cautionary Tale: Pregnancy and Face Recognition

3.4.1 Targeted Advertising for Pregnant Women

Loyalty cards are known to be used for better targeting of coupons and other marketing offers to customers of retail stores. With such cards, a retailer is able to record what products a customer bought at each visit at the store. Data scientists can use such large datasets for customer segmentation, mining association rules describing the products that are frequently bought together, to predict interest in a certain product (group) [88, 364, 62], and even to predict personal information about the customer. When customers sign up for such a loyalty card, these uses are typically communicated to the customer.

A story published by the *New York Times* in 2012 [118] reports the story of a large retail chain in the US, using such loyalty card data in an attempt to predict pregnancy [206]. The business motivation of such an exercise is clear: once someone has a baby, it is hard to change buying habits, and brand preference

---

[3] This is a story, loosely based on a similar 2007 tale of a user whose purchase of a diamond ring was posted on his Facebook wall without him knowing about it, revealing the surprise to all his Facebook friends, including his wife [249]. The Beacon program that enabled this was terminated in 2009 [69].

for baby products like nappies are setting in. Also, at that time, a customer is already overwhelmed with other marketing campaigns aimed at new parents. So how can you set up a historical dataset, with labels indicating pregnancy? One way is to use a baby shower registry, which reveals when the baby was born. This is what the retailer reportedly did [118]. Another way is to simply work backwards from the first time someone starts buying baby products, such as nappies, as roughly 9 to 10 months before the time the woman was likely pregnant.[4] Next, predictive modelling can be used where each product (type) that the retailer sells can be an input variable, having the algorithm figure out which ones are predictive for the stated target variable. And apparently, some changing patterns in buying behaviour seem to exist when being pregnant. The *New York Times* article describes that products as unscented lotions, scent-free soap and supplements like calcium and magnesium are products that pregnant women tend to buy [118]. These findings can then be used to score all customers and predict which ones are likely pregnant.

Assuming the model is accurate, and that the customers agreed to targeted advertising campaigns when signing up for a loyalty card, one can wonder what the issue is. The anecdote goes that at some point after the predictive campaign has been put into production, a man walked into one of the stores to complain: he was getting coupons in the mail for baby clothes and cribs, addressed to his daughter, who was still in high school ... The manager apologized and called the father back a few days later. The man reportedly stated [118]: 'I had a talk with my daughter ... It turns out there's been some activities in my house I haven't been completely aware of. She's due in August. I owe you an apology.'

Because of bad reactions received to such coupons of baby-related products, targeted at pregnant women, an executive at the retailer reportedly said that they started mixing the ads with things they knew pregnant women would never buy, such as a lawn mower, next to the baby product ads, in order to make the targeted advertising less obvious [118].

Although this case was heavily discussed in numerous newspapers and on-line blogs, some question whether the anecdote actually happened [150] and in a statement the retail company reportedly replied that the *New York Times* article contained inaccurate information [118]. In any case, it provides a cautionary tale that describes the ethical considerations that should be made in deciding the target variable.

---

[4] Of course, assuming the customer did not buy nappies at other locations previously. This target variable definition would surely include some wrong labels, but might work to find the general patterns of pregnancy.

As an afterthought, note that in this application the actual identity of the customer is not necessarily needed. As long as the customer keeps using the same loyalty card, linked to a unique customer ID, such analyses can be conducted. Not having a name, email, or address does imply that the targeted advertising can only happen in-store: the retailer could simply print the coupons on the receipt. This could even be done without any identifier: simply look at the products that were bought during a single visit. If the customer bought unscented lotions, scent-free soap, and calcium supplements, a near real-time prediction could be made that the current customer is potentially pregnant and should be given a coupon for nappies. This is an example of what Barocas and Nissenbaum describe as being 'reachable' while also being anonymous [38], as discussed earlier in Chapter 2, and clearly still demands ethical data science considerations.

Discussion 6

When it comes to predicting pregnancy, consider the following questions:

1. What 'likes' of Facebook pages would be predictive for pregnancy?
2. Would you be offended if you got an online ad (or offline coupon) for baby products, in case you or your partner were actually pregnant?
3. What if you would get the ad while not being pregnant? Under what circumstances would you be offended, or can you imagine other persons being offended?
4. What exactly would be considered unethical about targeted advertising for pregnant women? Is it because something is predicted before the customer or his or her friends know about this themselves? Is it because it is something physical? Is it because it is unexpected that a retailer would know about this? Any other reason?
5. What products, when bought by a pregnant woman, could be predictive for the gender of the upcoming baby?
6. Suppose you work at a large baby product manufacturing company, and your manager asks you to set up an online campaign, aimed at pregnant women. How would you respond?
7. For what companies or organisations would a pregnancy prediction model be ethical according to you?

8. For what other types of products do you feel that targeted advertising campaigns should not be used? Or put in data science terms: what target variables should not be used in predictive modelling for targeted advertising?

3.4.2 Face Recognition

Face recognition technology uses data science software to find a match for an image of a face, in a dataset of images with faces. It is one of these data science applications that comes with ample ethical concerns, due to the sensitive nature of biometric data, but also because of the wide potential use and misuse in our lives.

The advocates of such technology argue that it can solve crimes, assist in finding missing persons and prevent identity theft [67, 403]. The London Metropolitan Police Service is reported to already have deployed a face recognition system in 2001 [67]. In New York, the police department was able to solve a stranger-on-stranger shooting, where there was no DNA evidence, no fingerprints, and no eyewitnesses [308]. Many similar stories have been published. More recently for example, in 2020, the Toronto police department used face recognition software to solve a murder by comparing the image of the suspected shooter with a database of 1.2 million mug shots [355]. Not only police departments make use of face recognition, also international airports, casinos, and even construction job sites have been reported to do so [67]. In the US, the department of motor vehicles scans drivers' faces to prevent licence duplications and fraud [403]. A more humorous application of face recognition is the Google 'Art Selfie' app that matches a user's picture with a painting [174].

One of the events that ignited heavy discussions on the ethics of face recognition was the Super Bowl XXXV, in 2001. Tampa police used a face recognition system to reportedly scan tens of thousands of fans, in search of wanted criminals [403]. Despite trying to match the thousands of faces, only a handful of petty criminals were reported to be identified, though no one was detained. Journalists named this Super Bowl edition the 'Snooper Bowl' [403, 67]. What is actually potentially wrong with such technology?

Mainly privacy concerns are raised in such discussions. In the Super Bowl case, even though several measures were taken to safeguard privacy, still privacy concerns arose. Signs were apparently put up which informed the fans that a face recognition system was in use [67]. Also, the images that were being stored in the large dataset only included known offenders, while each image of the filmed fans was only used to compare with this dataset. If no match was found, the image was immediately discarded. In spite of these important safeguards, the commotion focuses on three concerns: privacy, error, and function creep [67]. Notice that these coincide with the GDPR article 5 concepts of lawfulness, accuracy, and purpose limitation, as we've discussed in Section 2.2.1. The reader is encouraged to reflect on the importance of the other article 5 principles related to data minimization, storage limitation, integrity, and confidentiality.

The main privacy concern is using technology to identify someone without having provided consent, or even being aware of the technology use. Persons are simply unable to turn off their face [403], as would be possible with online tracking of cookies or location services on smartphone. But do people really expect privacy in such public spaces, where TV stations are present and the event is broadcasted live, while fans are taking pictures? Nissenbaum argues that even though there is indeed a diminished expectation to privacy in public places, still justifiable privacy expectations remain [198], and hence face recognition violates the privacy. She motivates this reasoning by indicating that many persons feel dismayed when they learn about the surveillance practices in public places. The use of captured images of your face by TV cameras in a different context, requires different privacy guarantees. Brey argues that most visitors would likely not agree to be part of a police lineup to identity a suspect, even though they might possibly acknowledge being on camera. Since the context is different than the context of potentially appearing on television or in the background of a fan's picture at the Super Bowl, privacy might be violated [67].

A second ethical concern is the mistakes that predictive models are bound to make. It is mainly the false positives that are concerning: being questioned by the police because of wrongly being matched with a suspected criminal. But as Brey states, this is part of a balancing act between the number of good citizens that are mistakenly stopped and the number of suspected criminals that are caught [67], a trade-off well-known to data scientists to determine the final cutoff of the prediction score to make decisions. If the False Positive versus False Negative ratio is limited, the system's misclassification can be regarded as just part of operational workings, just as human officers would have in tracking suspects. A large ratio would argue that the system should be scrapped. In Tampa, the face recognition system was suspended in 2003, as it had not yielded a single arrest or positive identification [67].

Face recognition technology is rapidly evolving, and any system requires proper evaluation. A 2019 independent evaluation of the London police's trial of face recognition software, for example, revealed that of the 42 proposed matches, only eight were confirmed to be correct (wanted violent criminals) [278, 108]. A 14-year-old black schoolboy was reported to be among the wrongly identified and was fingerprinted because of it. In another 2018 evaluation of Amazon's face recognition software 'Rekognition', the American Civil Liberties Union (ACLU) reportedly found that 28 members of US Congress were misidentified as having been arrested for crimes, with a disproportional misidentification rate for people of colour [406, 266]. Academic research has similarly shown that some commercial gender classification systems suffer from substantial disparities in the accuracies across skin colour [73]. These are of course small and specific test runs conducted on software that might be obsolete as you read this, as face recognition software continuously is being improved. Yet, it shows that one should be aware of the limitations of such technology.

Another issue is 'function creep' [469, 67]. Function creep occurs when the technology is used for purposes that it was initially not designed for. Think of enlarging the dataset to also include images of political activists, journalists, citizens, etc; or the use of the software to track individuals over a longer time period, retro-actively being able to see where someone has been; or to allow access to the system by other organizations and persons, such as companies or individual operators.

These issues might be thought of in the context of the company Clearview AI [209, 207, 80, 354]. This American company provides face recognition software for law enforcement. With its app, you can take a picture of someone, upload it to their servers and it will return a set of publicly available pictures of the identified person, together with links to where the pictures appear online. The data it relies on is reported to have been scraped from Facebook, Twitter, Instagram, Youtube, and 'millions of other websites' [207], and reportedly contains over three billion images [354, 80, 207]. Clearview AI's CEO argues that it only scrapes images that are publicly available [354, 207]. This access to billions of images is a major advantage compared to the data available to law enforcement agencies, whose databases historically have been limited to government-provided images, such as pictures from driver's licences and mug shots [207]. The potential uses of such a system are security-related, as law enforcement can be helped in identifying suspects and potential terrorists. A story by the Indiana State Police describes how they were able to identify a shooter from video footage, for whom no mug shots or driver's licence was found [354]. So only a system that had data beyond those data sources, and proper face similarity matching software, would have been able to identify this person. Also businesses could benefit from the app, as it can help with identifying perpetrators of crimes as shoplifting and credit card fraud. Yet, clearly ethical issues arise with regards to data gathering and deployment. First of all, public data does not imply that the data is free-to-copy, as we discussed in Section 2.2. European regulations allow for a 'database right', where you are not simply allowed to copy or extract substantial parts of a database without the owner's consent, even if the content itself is not copyright protected. Also policies of individual websites (or other data sources) often tell you what you are allowed to do. For example, Facebook disallows all automated scraping except for a couple of listed companies but also limited to certain webpages, as revealed by www.facebook.com/robots.txt. So Clearview AI seems to have violated the policy by crawling Facebook pages, which resulted in Facebook and other companies such as LinkedIn, Twitter, and Youtube, demanding that Clearview AI stop the scraping [354].

Additionally, there are serious questions on who should be allowed to use such software. Should there be limitations to who gets to use the data: law enforcement agencies from all countries, security departments from businesses, or even private persons? A story goes that a potential investor of Clearview AI was able to take a picture of the date of his daughter that he happened to encounter in a restaurant, and was able to identify the person and his background through the app [209]. This seems to violate our privacy expectation. One can quickly imagine other potentially unethical uses of the app. The *New York Times* suggests three: identifying activists at a protest; stalking an attractive stranger who someone notices on the subway; or a foreign government digging up secrets from citizens and blackmailing them [207].

As you notice, once more a trade-off is discovered between utility of data (and data science) and privacy. Once it is established that the data and technology is legally acquired, that the results are accurate and can be of great value for law enforcement, one could argue that the use thereof by well-defined roles, and oversight, might be warranted in countries and organisations that can be trusted or audited to follow these set guidelines. Others might say that this is simply too big a privacy violation, and that the use of face recognition for mass surveillance boils down to 'basically robbing everyone of their anonymity', as stated by Joseph J. Atick, a pioneer in face recognition technology, in an interview with the *New York Times* [403].

This debate is clearly still raging in our societies, with both regulators and technology companies looking for the right balancing act. That this ethical debate will continue in the future is well reflected in this 2020 answer by Margrethe Vestager, the Vice President of the European Commission, responsible for competition and digital industry, to the question: Should mass facial recognition be prohibited? 'I do not know yet. It is a very sensitive issue. It may be useful, but we should not rush. We need to define precisely the conditions of its possible use, its virtues and its limitations. Under no circumstances should they impede freedom of expression or assembly, as reported in Hong Kong. Some American states have also given it up after trying it, claiming that it is not yet technologically reliable enough.' [259] In response to the growing societal concerns, and a lack of clear regulation on the matter, in 2021 Facebook chose to shut down their face recognition system and delete the 1 billion+ faceprints [347].

Discussion 7

Suppose you just installed an intelligent video doorbell. The accompanying app has a face recognition feature, where you can link a face to a person. This allows personalizing the messages and ring sounds you get: whenever the recognized person rings the bell, a special ring sound is made, or only recognized faces lead to a message being sent to your phone.

1. Do you think someone could mind that their face is linked to their name by your doorbell app?
2. How could the information be misused, by the doorbell company or by others who get unauthorized access to this dataset of faces with names?
3. Reflect on how the GDPR principles of article 5.1 of informed consent, accuracy, purpose limitation, and data limitation could or should be implemented in this setting.
4. Would you turn on this feature on your intelligent video doorbell?
5. What companies or organizations would you feel comfortable with, having a picture of your face and your name (and the necessary face recognition software to match another image of your face with your name)?

3.5 Fair Relabelling

Let's now turn to the discrimination aspect of fairness. How can we measure that a historical dataset contains a bias against a sensitive group, and how can we change the dataset in a preprocessing step so as to remove this measured bias from the dataset?

3.5.1 Measuring Fairness of a Dataset

To get rid of discrimination in a dataset, we must first define how to measure the extent of such discrimination or bias against a sensitive group in a dataset. Suppose we have a dataset with a certain sensitive attribute, called S, which takes value s for the sensitive group and ns for the non-sensitive group. For example, in case of discrimination against women, $S = s$ would indicate gender being female and $S = ns$ would indicate gender being male. The target variable Y to predict has a positive ($Y = +$) and negative ($Y = -$) outcome, for example being granted or denied credit, or being hired or rejected for a job. The first definition is statistical parity or dependence [233], defined by (3.1). This measure takes the probability for a positive outcome for the non-sensitive and sensitive groups, and substracts them. The second definition of disparate impact, defined by (3.2), takes the ratio of these two numbers. These probabilities can easily be calculated from the data: for example the probability for a positive outcome, given the instance is part of the non-sensitive group, can be measured as the number of instances with the non-sensitive attribute value that also have a positive outcome (ns^+) divided by the total number of instances with a non-sensitive attribute value (ns^T).

$$\text{Statistical parity(D)} = P(Y = +|S = ns) - P(Y = +|S = s) \quad (3.1)$$

$$= \frac{ns^+}{ns^T} - \frac{s^+}{s^T}.$$

$$\text{Disparate impact(D)} = P(Y = +|S = ns)/P(Y = +|S = s) \quad (3.2)$$

$$= \frac{ns^+}{ns^T}/\frac{s^+}{s^T}.$$

Consider the example dataset of Table 3.8, similar to the one used by Kamiran and Calders [233]. As calculated below, the statistical parity is 40%, meaning that men have a 40% higher probability than woman to be hired, in absolute terms, while the disparate impact is two, which means that men are twice as likely to be hired than women, in relative terms.

Table 3.8 Example dataset with 5 men ($S = ns$) and 5 women ($S = s$), with in total 6 positive outcomes ($Y = +$) of being hired.

id	...	S	Y
1		ns	+
2		ns	+
3		ns	+
4		ns	+
5		ns	−
6		s	+
7		s	+
8		s	−
9		s	−
10		s	−

$$\text{Statistical parity(D)} = \frac{4}{5} - \frac{2}{5}$$
$$= 0.8 - 0.4 = 0.4.$$
$$\text{Disparate impact(D)} = \frac{4}{5} / \frac{2}{5}$$
$$= 0.8/0.4 = 2.$$

3.5.2 Massaging

The problem we need to address is that the historical dataset might exhibit discrimination against a sensitive group, for example women being hired less likely. This could be resolved by relabelling the data, by changing the class label of some instances from the sensitive group from negative to positive, and from positive to negative for some instances from the non-sensitive group. In our example, this implies changing the label for some women from not hired to hired, and for some men from hired to not hired. This approach has been put forward by Kamiran and Calders, which they coin as 'Massaging' [233], and is briefly discussed next. Two design questions arise: how many instances need to be relabelled, and which ones?

First: how many? Well, as many as needed to make the measured discrimination become (close to) zero. So if we relabel M men that were hired and M women that were not hired, the measured discrimination, in terms of statistical parity of our dataset, becomes:

$$
\begin{aligned}
disc'(D) \quad &= \quad \frac{ns^+ - M}{ns^T} - \frac{s^+ + M}{s^T} \\[2mm]
&= \quad disc(D) - M \cdot \left(\frac{1}{s^T} + \frac{1}{ns^T} \right) \\[2mm]
&= \quad disc(D) - \left(M \cdot \frac{s^T + ns^T}{s^T \cdot ns^T} \right) \\[2mm]
&\equiv \quad 0
\end{aligned}
$$

If we bring M to one side, we get the required number of changes for each group: the measured discrimination in the dataset, multiplied with the number of persons in the sensitive group (s^T), and the number of persons in the non-sensitive group (ns^T), divided by the total size of the dataset (the sum of the number of persons in the sensitive and non-sensitive groups, $s^T + ns^T$).

$$
M = \frac{disc(D) \cdot s^T \cdot ns^T}{s^T + ns^T}. \tag{3.3}
$$

Returning to our example dataset, we find that we need to change the label of one man (who was hired) and one woman (who was not hired).

$$
M = \frac{0.4 \cdot 5 \cdot 5}{10} = 1. \tag{3.4}
$$

That solves the first design question, but now the question arises: which instances to relabel? For that, Kamiran and Calders argue to use a scoring classifier built on the original dataset to score all data instances. Which man would we want to change from positive to negative? The one with the lowest predicted probability to be hired, because for that person we are the least certain about whether to invite him or not. Similarly, which woman do we want to promote from not being hired, to being hired? The one with the highest probability to be hired (or the lowest probability to be rejected). By doing this massaging, a dataset is obtained where the measured discrimination is (close to) zero, while at the same time not having changed the dataset too much, so as to stay close to the original dataset and ensure that the model is still accurate when used in real-life settings.

The authors of this 'Massaging' paper also propose two other preprocessing methods, based on weighing each instance according to how much the observed and expected probabilities to be hired are different, whether you are part of the sensitive group or not [233]. These weights can then be used in classification algorithms that are able to work with weighted instances (in a method termed 'reweighing'), or can be used to resample the dataset (in a method termed 'sampling').

3.6 Cautionary Tale: Biased Language

Unfair data science goes beyond predictive modelling. The following cautionary tale shows how data representation learning can also include bias. 'Man is to Computer Programmer as Woman is to Homemaker?' is how Bolukbasi et al. start their influential 2016 paper [59]. They applied a commonly used word embeddings approach to Google News data, and reveal startling gender biases in news stories.

Computers can only reason with numbers, so the text needs to be transformed to a numerical representation. A powerful and widely popular approach to do so is word embeddings [229]. The idea is to come up with a vector representation for each word, such that words that are similar to one another will be close to each other in this numerical vector representation. Similar words can be seen as words that are often used with the same other words before and after it in a sentence. The words *Belgium* and *France* for example are often used in the same context, so they should lie close to each other. The different new numerical dimensions can be seen as concepts that help to define the meaning of a word. For example, one dimension might encapsulate the concept of royalty, another might be more related to country.[5] In case of a traditional 'bag of words' encoding, the words *king* and *queen* would be as different as the words *king* and *data*. In the embedded representation this would no longer be the case: *king* and *queen* often occur in the same context/sentences (e.g. 'The king wears the crown.' and 'The queen wears the crown.') and hence will be closer to each other in the new embedded dimensions than the words *king* and *data*.

Word embeddings facilitate reasoning with words:[6] what if I take the word *king*, subtract the word *man*, and add the word *woman*? This idea is illustrated in Figure 3.2, and shows that you would end up somewhere very close to the word *queen*, as we would expect. Now what happens if we similarly look at the female version of the word *computer programmer*? As Bolukbasi

---

[5] These dimensions don't necessarily have an explicit meaning, and are determined fully data-driven.

[6] When we reason in the embedded space, a word corresponds to a vector, and calculations with words hence correspond to calculations with vectors.

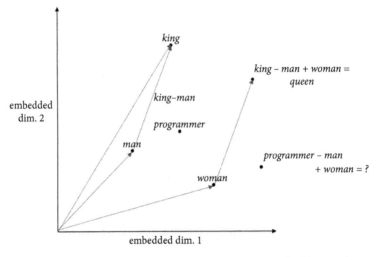

Fig. 3.2 Reasoning with word vectors using word embeddings. The female version of the word vector *king* is *queen*, as derived from vector additions: *king - man + woman*. The female version of the word *programmer* turns out to be *homemaker*.

et al. already reveal in the title of their paper, we end up with the word *homemaker* [59]. This language bias is shown to exist in the word embedding with 300 dimensions that one obtains from analysing Google News articles, with 3 million English words. And it is far from the only example of our biased language use. Some other stereotype male/female analogies that are found include [59]: nurse/surgeon, housewife/shopkeeper, interior designer/architect, and diva/superstar.

These analyses reveal a bias in the language of news articles, which we can assume is written (largely) by professional journalists. And so we find once more that not being aware of this bias will lead to biased models that are built upon them. In this specific context, a major concern is the widespread use of word embeddings, with hidden gender bias, which might be amplified in the resulting applications, be it resume screening or online querying [59]. Yet another cautionary tale that even if the source of the text might seem trusted, be aware of hidden biases and their implications of conducting data science on the data.

A related and interesting online visualization tool can be found on the website of the University of Maryland, and is the result of a 'hacking discrimination' hackaton held at Microsoft in 2017 [119]. The tool allows you to look for stereotypes yourself. You can look at, and play around with, visualizations of male and female adjectives, drinks, and many more categories. One remarkable result is that the most 'male' adjectives include *certainly*, *coky* and *decent*, while the most 'female' adjectives are *sassy*, *sexy*, and *gorgeous*.

The solution would rather obviously exist in debiasing the learned representation, by ensuring that words that are considered to be gender neutral (such as *surgeon*) have the same distance to each word of gender pairs, such as *he* or *she* [59]. At the same time, the embeddings should correctly maintain expected gender analogies, for example requiring that the word *mother* lies closer to the word *she* than to the word *he*.

Language bias has not only been found in Google News articles. Claudia Wagner et al. looked at potential gender bias in Wikipedia articles [457], another source of text that one would expect to be neutral in terms of gender. Wagner and her co-authors found that Wikipedia articles about women more often discuss the gender, romantic relationships, and family-related topics (such as her husband, his job, and kids) than is the case for Wikipedia articles about men. Similarly, they found that some words are predictive for the gender of who the article was about: words such as *baseball* and *football* are found to be discriminative for articles about men, and words such as *husband* and *divorced* for articles about women [457].

3.7 Summary

This chapter considered how to measure and ensure fairness of your dataset, in terms of privacy and discrimination against sensitive groups. Too often fairness is assumed by simply not using personal identifiers or sensitive attributes. But this is like simply closing your eyes as a solution for a flat tyre on your car: you might not see the problem any more, but it is surely still there.

The privacy of (preprocessed) datasets can be defined in terms of k-anonymity, l-diversity, and t-closeness, where each attempts to counter certain privacy attacks. Such attacks aim to re-identify a person or reveal the value of a sensitive attribute for a person, and do not necessarily require being able to link a specific data point to a person (cf. homogeneity attack). The methods to ensure such privacy are suppression of values and grouping of instances or values.

Keep in mind that satisfying these privacy definitions does not mean there is total privacy and utility: the continuum between privacy and utility is

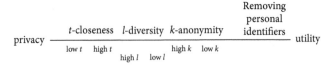

Fig. 3.3 Trade-off between utility for the data, and privacy.

illustrated in Figure 3.3.[7] Stronger definitions, or larger k's, l's, or smaller t's, lead to more privacy, at the expense of utility, as the dataset is made more general. So be careful when you call the dataset anonymous, and be precise in what definition you use. If there are still circumstances in which a person could be re-identified, the data is better called pseudononymous, as defined in GDPR.

Behavioural data poses specific privacy risks, mainly due to the fact that very few persons have the exact same recorded behaviour. This is the driving force behind re-identification attacks, linking for example Netflix data with IMDb data. Imposing k-anonymity strongly reduces the utility of such data, as the fine-grained property of such data is what typically leads to a high accuracy for the predictive models built on such data. The different case studies on movie viewing, online search and location data show the serious privacy risks that may come from publishing this data, even if the motivation for making the data public is well-intended.

The mentioned privacy attacks assume that adversary Adrek has background knowledge. In some cases this might be far-fetched, but reveals how tricky it is to publish personal data. In the next chapter, we'll show that by adding noise to the result in an intelligent matter, more formal, mathematical privacy guarantees can be provided even in the presence of background knowledge, future technologies, or unlimited computing resources.

The other fairness issue we discussed was discrimination against sensitive groups, firstly, by looking at the target variable. It is important to properly define this and consider historical biases that might creep into this definition. Think for example of historical preferences for men in hiring or promoting. Additionally, there should be ethical limitations on what you predict. Even though the data on some target variable is available, predictions are accurate, and consent was provided, does not mean that you should predict that variable. The pregnancy case is a cautionary tale for this point. Deciding what is ethical

---

[7] When interpreting Figure 3.3 note that the shown order is not always total: it is possible that for a dataset having l-diversity with some low l is worse in terms of privacy than some k-anonymous dataset with a high k.

to predict once more depends on the context and application. Predicting pregnancy might be considered ethically questionable for marketing purposes; it will have much less concerns in a medical setting.

Let's briefly revisit the opening story of the chapter now, where a university is asked to disclose for each student the courses he or she is enrolled with, the grades, days absent due to COVID-19, nationality, date of birth, postal code, and gender. The request from the minister of education wrongly assumes the data is anonymized because the names are hashed. The hashing rather leads to pseudonymization. Indeed, Sweeney demonstrated that the combination of date of birth, gender, and zip code will likely identify many students uniquely [422], and hence could reveal sensitive information about them, such as their grades or COVID-19 diagnosis. Including more privacy through k-anonymity and differential privacy surely seems warranted in this case.

Finally, we looked at how to measure and remove potential discrimination against sensitive groups in a dataset. We discussed two important measures for dataset fairness: statistical parity and disparate impact. To remove historical bias in a dataset, one can either work on the target variable, by changing the target variable to a positive outcome for some of the negatively discriminated sensitive groups, and vice versa; or by working on the instance level, by providing more weight or oversampling these negatively affected persons, while providing less weight to or undersampling the other groups. An advantage of removing the bias from the historical dataset is that many techniques and analyses can be applied afterward. Ensuring fairness in the modelling phase will make the approach technique specific.

4

Ethical Modelling

In November 2019, tech entrepreneur David Hansson wrote a series of Tweets that went viral, in which he wanted to address an apparent injustice done to his wife [5, 318]. Apple Card offered him 20 times the credit limit that it offered his wife, although they have shared assets and she has a higher credit score: '[My wife] spoke to two Apple reps. ... The first was like "I don't know why, but I swear we're not discriminating, IT'S JUST THE ALGORITHM". [5] Apple co-founder Steve Wozniak replied 'The same thing happened to us. I got 10× the credit limit. We have no separate bank or credit card accounts or any separate assets. Hard to get to a human for a correction though. It's big tech in 2019.' [471] This media storm led to a formal investigation by the New York Department of Financial Services into the potential sexist credit decision making by Goldman Sachs, who was responsible for the credit scoring for the Apple Card [5, 318]. This story includes several components that will be addressed in this chapter. First, the issue of implied discrimination against women. We'll discover how we can assess discrimination against a sensitive group, with various measures. Surprisingly, these measures often conflict, so motivating the choice of the selected fairness measure is important, as will be demonstrated with a case from predicting recidivism. A second issue that is foundational for this tale is explainability: being able to explain predictions. This need stems from various sources: to obtain trust in the model, but also to get insight, and to improve the model. Several techniques will be introduced, ranging from LIME and counterfactuals that lead to individual explanations (which would be well suited in this credit scoring context and arguably what both David Hansson and Steve Wozniak are asking for), to rule extraction methods that explain the global prediction model. But first, this chapter will discuss various techniques that attempt to reconcile privacy and data science modelling with personal data.

Data Science Ethics. David Martens, Oxford University Press.
© David Martens (2022). DOI: 10.1093/oso/9780192847263.003.0004

4.1 Privacy-Preserving Data Mining

4.1.1 ε-Differential Privacy

Defining Differential Privacy

One of the golden standards to include privacy in data science is the concept of 'differential privacy', which originated from Cynthia Dwork [122]. Once more the question is asked, how can we analyse the data while still preserving privacy? In a previous chapter this concept was already introduced in a decentralized form, by adding noise before recording the answer. In the centralized version that we will discuss next, the noise is added to the result before communicating it to the outside observer. In other words, the data analyst is trusted, but the outside observer is not. Think for example of census data that is collected by the government. These can be used in various valuable data science exercises or to assign government resources to different regions, but can also reveal very personal and sensitive information. The goal of differential privacy is to allow a social scientist to share useful statistics about sensitive datasets. For example: How many people in Belgium have HIV? How many students receive financial aid? Or how many 'Data Science Ethics' students pass the exam? In itself these are all interesting questions to ask, but not accounting for privacy might reveal personal and sensitive information.

Let's first consider again what the privacy issue would be when reporting simple counts. As we just saw with k-anonymity, background knowledge can lead to personal information being revealed. Here is an example, similar to the one described by Wood et al. [467]. Suppose in a study for a course on *Data Science Ethics*, the professor reports that in March 2020, there were 50 students taking the course, 10 of which were on financial aid. This statement in itself does not reveal personal information. Now suppose the month after, the TA (teaching assistant) reported the same information, stating: in April 2020, there were 49 students taking the course, 9 of whom were on financial aid. Once more, this is not revealing personal information in itself. But when we combine both statements with background information, troubles emerge. If student Sam knew that all 49 students also participated in the March 2020 course and that student Tim dropped out of the class in March, then Sam would now know that student Tim was on financial aid. And hence Tim's personal information is being revealed to Sam by this simple counting statement. Similarly, when the government reports daily the cumulative number of COVID-19 cases per city, a sudden drop of one and the background information that a person moved out of the city, might reveal that this person was diagnosed with COVID-19. That is what differential privacy aims to avoid.

Now let's see how to define privacy of an algorithm. The basic idea is as follows: whether an individual's data is used or not by the algorithm, the outcome should be essentially the same. What 'essentially the same' means is formally defined by a ε parameter, as shown next [122, 213, 467]:

ε-**differential privacy** [122, 467]: 'A property of an algorithm where the outcome or result will remain **essentially the same** whether you participate or not in the dataset.'
More formally: 'if for any two datasets D and D' that differ only in 1 data instance, and for all possible outcomes S from algorithm A, and $\varepsilon \geq 0$:
$P(A(D) \in S) \leq e^{\varepsilon} \cdot P(A(D') \in S)$'

This definition begs for some examples to get a more intuitive interpretation. In the example on p. 122 (inspired by the work of Wood et al. [467]), whether you are in the dataset of the city reporting the number of COVID-19 cases or not, or whether you have dropped out of the course (and the dataset) or not, the reported numbers should not differ much and the impact on your privacy, which can now be quantified in terms of ε, should be small. Suppose the same professor asks her students to state whether they like the course or not. A student might fear that participating in this survey and stating she dislikes the course could lead to a more difficult exam for her. Differential privacy ensures that the probability that the exam will be more difficult, would be roughly the same whether she participates or not. For a small ε, $e^{\varepsilon} \approx 1 + \varepsilon$, so mathematically for $\varepsilon = 0.01$ this would become: the probability that the exam will be more difficult when NOT participating in the survey should be less or equal to 1.01 times the probability that the exam will be more difficult exam when participating in the survey. So if the probability that the exam became more difficult after the survey is 1% without participating, it should increase to maximum 1.01% when participating. So notice that it doesn't mean that the exam would not be more difficult, just that participating or not doesn't change this much.

The ε privacy parameter as a measure of privacy loss

The ε parameter is called the privacy loss parameter [467]. The smaller the ε, the more privacy is guaranteed, as the change in probability to any outcome with or without an individual's data becomes smaller and smaller. What would happen if we set ε to zero? Well, then these probabilities need to be exactly the same, which can only occur if there is only noise, and no signal, in the data. So total privacy and zero utility is obtained for $\varepsilon = 0$, with larger ε values leading to less privacy guarantees.

Remember that differential privacy is achieved by adding noise to the outcome of the analysis. The definition now provides us a mathematical method to do so. Suppose we want to count how often the binary variable x_i (with $i = 1, 2, ..., n$) equals to 1, hence take the sum of this variable. One use case would be to calculate how many citizens have COVID-19 or how many students like the course. Then the following algorithm, which adds Laplace distributed noise with zero mean and a standard deviation of $1/\varepsilon$ to the actual sum, is shown to be ε-indistinguishable[1] [122, 467]:

$$A(x_1, x_2, ..., x_n) = \sum_{i=1}^{n} x_i + \text{noise}(\varepsilon), \text{ where } \text{noise}(\varepsilon) \sim Lap(1/\varepsilon). \quad (4.1)$$

Figure 4.1 shows the Laplace probability density function for two different ε values. The number on the X axis is the possible noise that is added; the number on the Y axis is the probability that the noise on the X axis is added. As you notice, it favours small values around zero, much more so than a normal distribution. The larger the variance (the smaller the ε), the more likely that noise is being added, hence the greater the privacy guarantees. Returning to our example, instead of reporting that in March 2020, 10 students which were on financial aid, we'd say $10 + \text{noise}(\varepsilon) = 12.4$ students (or approximately 12) were on financial aid.

Setting a proper ε is unfortunately not formally defined and depends on the trade-off between privacy and accuracy that one wants to obtain. As a rule of thumb, values between 0.001 and 1 tend to be chosen [328]. Consider a small dataset of 10 records, and the simple sum function. Adding the noise from the distribution with $\varepsilon = 0.01$ in Figure 4.1 will lead to very inaccurate estimates of the sum, where the noise accounts for most of the answer. For a very large dataset with millions of records, the effect on the accuracy of the answer is smaller. Differential privacy will hence require increasing the minimal dataset size needed to provide accurate results. This leads to another rule of thumb which states that 'almost no utility is expected from datasets containing $1/\varepsilon$

---

[1] Other, more complex noise introduction methods have been researched and proposed as well [122, 213].

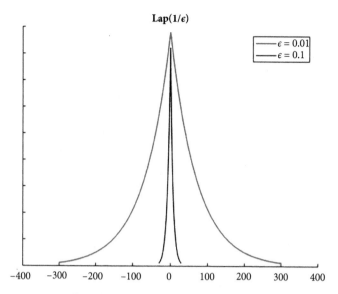

Fig. 4.1 Laplace distribution with mean 0 and standard deviation $1/\varepsilon$.

or fewer records.' [328] So for a desired ε of 0.01 this implies that the dataset should be of size 100 or more. Optimal ε values are still the focus of current research, where some research looks at economic cost to set ε [213], while others will set ε based on the required accuracy, size of the dataset and some other parameters [328].

So far, we've seen the use of differential privacy for single count queries: how many persons have a disease or like a course. What if we combine different analyses? If we answer the same question over and over again, the average answer will go to the real answer, as the noise has zero mean. In other words: the privacy risk increases as we do more analyses. Differential privacy is compositional, where the privacy leakage accumulates in an elegant manner over more analyses [467, 213]. The combination of two differentially private studies, with respectively ε_1 and ε_2 will be differentially private, with $\varepsilon = \varepsilon_1 + \varepsilon_2$. This ε can therefore be seen as a privacy budget [467]: if you want to report the results a total of k times, and have an overall privacy budget of ε, then each study itself must have a privacy parameter of ε/k. This privacy parameter reflects how much privacy budget an analysis can use, and how much the risk to an individual's privacy may increase. The combination of many simple differentially private analyses can lead to advanced analysis techniques, where differentially private versions of clustering, regression, and classification techniques have been proposed [328].

Promises and Uses of Differential Privacy

The mathematical guarantee given by the differential privacy definition holds, no matter the background knowledge [123, 467], available computing resources, or even available technology. In other words, it is future proof [177]. What it does not promise is that still some sensitive attributes can't be predicted for you based on the analysis. If a differentially private survey study looks at the average salary of a professor of age between 40 and 50 from Antwerp, whether I participate or not would not change the result much (as mathematically guaranteed). But the result itself will reveal a good estimate of my salary to everyone reading the results of the survey and knowing my profession and age.

A large-scale implementation of ε-differential privacy was used at the US Census Bureau for the 2020 Decennial Census [193]. The US Census Bureau counts each resident of the country, together with some socio-demographics and where they live on 1 April of every tenth year. The very first US Census study was conducted in 1790. This data is used to determine the number of seats for each state in the US House of Representatives and in applications such as economic analyses, the distribution of federal funding to states and communities, and emergency response plans [443]. An internal study revealed that for their Census 2010 data, 52 million individuals could have been re-identified based on external data, with their race and ethnicity being reconstructed [193]. The Census Bureau understood that the traditional disclose avoidance methods became insufficient for the available external databases and computing power, leading to the use of ε-differential privacy in the 2020 US Census.

In Chapter 2 the decentralized version was introduced, which we can now revisit. Remember that in that version, noise is added before the data is recorded, while the centralized version adds noise to the result. Our coin flipping example would record the correct answer in 75% of the cases. This corresponds to the mathematical ε-differential privacy guarantee with an ε of $ln(0.75/(1 - 0.75)) = ln(3) \approx 1.1$ [122]. If the correct answer is recorded in 99% of the cases, then $\varepsilon \approx 4.6$, meaning much less privacy guarantees as much less noise is being added. The centralized version requires you to trust the data analyst; the decentralized version does not require this. The choice between these versions depends on the risk of the recorded data to be hacked, subpoenaed, or being looked at inappropriately by internal employees at the data analyst [123]. If these risks are limited, the centralized version can be used (as is the case of the 2020 US Census); if the risks are larger, then the decentralized version can be used (as is the case of Google and Apple).

In summary, differential privacy provides future-proof privacy guarantees for an algorithm, by adding intelligent noise, with as main use case the privacy-friendly calculation of aggregate statistics. In what follows we'll consider some algorithms and protocols that are tailored for specific analyses and contexts, and continue to address this balancing act between performing accurate data science analyses while protecting privacy rights.

4.1.2 Zero-knowledge Proof

Understanding zero-knowledge proofs can lead to a true 'Eureka' moment. These protocols are focused on a specific kind of calculation: proving statements about a secret. A zero-knowledge proof is defined as follows:

Zero-knowledge proof [170, 167]: *A zero-knowledge proof is a method where one party proves a statement about a secret to another party, without revealing the secret.*

In the context of data science ethics, the secret is mostly personal data, for example, proving you are a citizen of the European Union, without revealing your nationality, or proving your income is higher than 2,000 Euro without revealing the exact amount, or proving you are over the age of 18, without revealing your age. But how can this work? A simple toy example to illustrate [330]: suppose you draw a card from a deck of 52 cards. The statement you want to prove is: 'The card I drew is red', while the secret is the exact card you're holding. How can you prove to a verifier that it is indeed red? A zero-knowledge proof is the protocol of simply going through the remainder of the deck and showing all 26 black cards. Since all black cards are revealed it is now proven that the card you have is red. Another popular example is based on the game: 'Where is Waldo?'[2] by illustrator Martin Handford [464]. In a large illustration, one needs to find a small figure named Waldo. The statement you want to prove is: 'I know where Waldo is on the illustration', while the secret is the exact location of Waldo [313]. A zero-knowledge proof involves a large board, double the width and double the height of the illustration, with a small cut-out in the middle, the size of Waldo. By placing the board on the illustration, with the cut-out revealing Waldo, you proved that you know where Waldo is (he is

---

[2] Known in the US as 'Where's Wally?'

revealed), but not showing where in the illustration he is (the board covers the rest of the illustration, including the edges of the illustration).

Zero-knowledge proofs find their origin in the 1980s at MIT [169] and have gained in popularity due to the ability to analyse personal data in the form of a proof, while protecting the privacy of the data subject.

A zero-knowledge proof needs to fulfill three criteria [167, 310]:

1. Completeness: if the statement is true, the protocol (followed by an honest prover) will *fully convince* the verifier.
2. Soundness: if the statement is false, a *cheating* prover will only convince an honest verifier that the statement is true with *some small probability*.
3. Zero-knowledge: if the statement is true, the verifier *learns nothing else* than that the statement is true (the secret is not revealed).

As an exercise, go over the zero-knowledge proofs just discussed and verify the three criteria. One more simple problem that can be solved with zero-knowledge proofs: proof to a verifier who is colour-blind, that you can distinguish between a green and red ball that the verifier is holding in its hand [84, 82]. The colour-blind person keeps one ball in the right hand, and one in the left hand. This verifier then places his hands behind its back and chooses whether to change the balls from hand or not. Next, the balls are shown to the prover, who needs to state whether the colour-blind verifier has changed the balls or not. Guessing this correctly the first time could be due to luck (a 50% chance), but by repeating the experiment n times, the probability that the prover is just guessing (cheating) and getting it right each time becomes $1/2^n$. So that covers the soundness property. At no time will the verifier know the colour of each ball (zero-knowledge), while also convincing the verifier that the prover knows the colour of the balls (completeness).

Especially interesting problems related to data science are zero-range proofs and zero-knowledge set proofs [77]. The zero-range proof aims to prove that a value is in a certain range, without revealing the exact value. This can be of great value for applications where model subjects need to reveal their income. A decision rule or tree might state that 'if income $> 2,000$ Euro and profession is data science then credit score = high'. With a zero-range proof, a model subject can prove his income is high enough without revealing the exact income. As such, the model subject is not giving away its data, yet can still be scored. Such a proof is also of great value for cryptocurrencies, to add privacy to everyday transactions [310, 211]. Bitcoins are in reality only pseudonominized. To protect the amount of a transaction, the sender and the receiver, zero-knowledge proofs can be used. Z-cash is an example of such a cryptocurrency that leverages the power of zero-knowledge proof [51, 211]. As with several of the privacy-technologies we're discussing, the computational requirements are (at the moment of writing this book) much higher [51]. Yet, the promise it holds, of privacy and personal data use at the same time, make it a technology that holds much promise for the years to come.

While zero-knowledge proofs provide a binary answer about a statement on protected data, the next technology, homomorphic encryption, allows calculations on protected data.

4.1.3 Homomorphic Encryption

Homomorphic encryption provides a solution to the combination of data protection and the increased popularity of cloud computing. As we've seen, encryption allows you to securely send personal data from a client to a server. But once we want to perform some calculations on that data, we will have to decrypt it before making the computations. At that time the data is no longer secure and revealed to the server, as illustrated in Figure 4.2. Some organizations might not trust (public) cloud computing for that exact reason: a bank might not want all personal financial transactions and information on its customers to be revealed to a cloud computing platform. Similarly, some citizens might not want their entire medical history to be shared by a hospital with a cloud computing service. And even if the cloud computing platform itself is

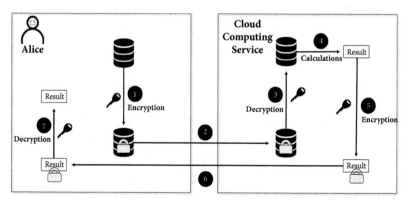

Fig. 4.2 When using a cloud computing service to perform calculations on Alice's dataset, the dataset will be on the server in plain text at the time of the calculations, and hence revealed to the cloud computing service.

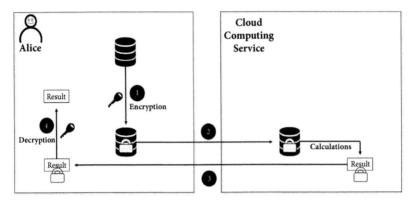

Fig. 4.3 By performing calculations directly on the encrypted dataset, with homomorphic encryption, the dataset is never revealed to the cloud computing service.

to be trusted with the personal data, one can still imagine that a data leak of the server at that moment would also reveal the personal data. That is where homomorphic encryption comes in. Homomorphic encryption is an encryption scheme where computations can be done directly on the encrypted data itself [163, 2], as illustrated in Figure 4.3. So a bank would be able to send all the personal data and financial transactions of its customers to the cloud, make use of their cloud computing services to daily calculate credit scores of their customers, and all this without ever having to decrypt the data. Or we would all be able to send Google search queries encrypted to Google servers, which would then return an encrypted result, which would then be decrypted on our computer [163]. In that case, Google would not know what we queried, just that we queried something, that Google made the search and returned the encrypted result.

This approach can be compared to a jeweller who has valuable raw materials such as gold and diamonds, and would like hired workers to make jewellery out of it [163]. The metaphor is that the raw materials are personal data, the hired workers are the cloud computing platform and the jewellery are the calculated results. The jeweller would place all the valuable raw materials in a box that she locks, and only she has the key to re-open it. The transparent box however has two openings with gloves attached to it, such that the workers can work on it through the gloves, yet cannot take any of the gold or diamond out of the box. Once the workers are done, the jeweler can take out the finished product. Homomorphic encryption is the encryption scheme which would correspond to such a locked box.

Unfortunately, we are not there yet, as 'fully homomorphic encryption' (FHE), where we can perform any function, be it multiplication, addition, search, or mean for an unlimited number of iterations, is not ready for practical applications yet. In 2009, in his seminal PhD work at Stanford, Craig Gentry described a theoretical way to perform fully homomorphic encryption [163]. Unfortunately, due to the major computational overhead, this approach still remains a theoretical exercise ten years later. If we want to perform data science calculations like deep learning or simple searches, we'll need this full homomorphic encryption to be computationally efficient. An additional advantage of the blueprint provided by Gentry is that the scheme is based on lattice encryption, and not the factoring of large numbers, which implies that the encryption would also be safe in a quantum computing era.

We did already see a 'partially homomorphic encryption' scheme [376, 163]: RSA. A partial homomorphic encryption (PHE) scheme only allows for one specific function to be performed on the encrypted data, in this case multiplication. Consider that we have two numbers m_1 and m_2 that we want to multiply on a server. We encrypt the numbers to c_1 and c_2 and send them to the server. As revealed by Equations (4.2) and (4.3), if we multiply the RSA encrypted numbers, send the result back to the customer, and decrypt that result, it turns out to be the same result as when we perform the multiplication directly on m_1 and m_2:

$$c_1 \cdot c_2 \quad = \quad (m_1^e \bmod n) \cdot (m_2^e \bmod n) \tag{4.2}$$

$$= \quad (m_1 \cdot m_2)^e \bmod n. \tag{4.3}$$

Homomorphic encryption would also be very beneficial for big data research. In many applications, the sharing of large datasets with personal and sensitive data, such as Facebook likes, payment data from banks, or online browsing data over a wide range of websites, is extremely limited. Even when such data can be worked on by data science researchers, it is often still limited to working on it on-site. Reproducibility of the application of (novel) data science algorithms on such data has so far been a difficult research issue. Similarly, simply the lack of access to such data holds off important research [178]. Fully homomorphic encryption would allow researchers to work directly on the encrypted data, without the need of any personal data to be revealed.

Because of the great promise of fully homomorphic encryption and the advances in computational speed, this is a technology that might not be ready for practical applications now, but would be of great value in the future for data protection.

In differential privacy, zero-knowledge proofs and homomorphic encryption, there are two parties: the one with a secret or personal data, and the one performing some analysis. The following methods involve situations where there are multiple parties, each having some personal data that they do not want to share with the other parties, yet where the use and analysis of the personal data of all parties would have benefits for all.

4.1.4 Secure Multi-Party Communication

There are many use cases where multiple parties can benefit from performing a joint data science analysis on their personal data, while keeping the data of each party secret to the others. As a running example, consider that banks wish to report what the average salary of a data scientist is in their sector, but they don't want to reveal the salary at their bank with the other banks. One approach to do so is with a trusted third party, an academic or governmental research group, for example, that receives the salary numbers from the different banks, calculates the average, and then reports this number to all banks. In that way, no bank will know what each of the other banks pays its data scientists, yet knows the average over all banks. What Secure Multi-Party Communication (SMPC) aims at is doing the same exact calculation while retaining privacy and without the trusted third party [136]. Other examples are the counting of votes per party, without revealing the vote of any single person, or an auction where individual bids are not exposed to the bidders and only the winning bid is revealed.

Introduced by Yao in 1982 [473], more formally we have m parties $P_1, P_2, ..., P_m$ with each some secret input $x_1, x_2, ..., x_m$, and we wish to calculate $f(x_1, x_2, ..., x_m)$ such that the result is accurate, the privacy preserved and without a need for a trusted third party. 'Secure communication' refers to the methods that performs data analysis without a trusted third party, while keeping the (personal) data secret. The term 'multi-party' of course refers to the many players with each having secret data. In the example of banks wanting to calculate the average on their data scientists' salaries, a SMPC protocol would go as follows.

Suppose we have three banks, each with a salary (we assume fixed) that they offer their data scientists. In a first step, each bank randomly divides the amount in 3 shares:

- Bank 1: Salary 1 = 2,000 = 545 + 1,050 + 405
- Bank 2: Salary 2 = 3,000 = 320 +   980 + 1,700
- Bank 3: Salary 3 = 4,000 = 888 + 1,011 + 2,101

In the next step, each bank keeps the first number, sends the second to the next bank, and the third number to the other bank. So Bank 1 would keep the number 545, receives the number 320 from Bank 2 and the number 888 from Bank 3. Based on these numbers, each bank calculates the partial average:

- Bank 1: Partial avg 1 = (545   +   320 + 888)   / 3 = 584.33
- Bank 2: Partial avg 2 = (1,050 +   980 + 1,011) / 3 = 1,013.67
- Bank 3: Partial avg 3 = (405   + 1,700 + 2,101) / 3 = 1,402.00

The banks send each other the partial results and take the sum, which results in the global average of the salaries: 584.33 + 1,013.67 + 1,402.00 = 3,000.00. We see that the calculation is indeed correct, while no bank knows what the other banks offer as salary to their data scientists, and without the need for any trusted third party.

A specific type of SMPC with various interesting applications is Private Set Intersection, where each party has a set of members where the parties want to calculate the intersection of these sets, while keeping all non-intersecting members secret [87]. Banks could for example be interested in knowing if any of their customers also have outstanding credit at other banks. Hospitals might similarly be interested in knowing if an incoming patient already has medical records at other hospitals. In a counter-terrorism setting, intelligence agencies from different countries might want to know if other agencies have files on a suspected terrorist as well. In each of these use cases, sharing all the data with a third party would add more privacy risk, while some players might simply not be willing to share this data. SMPC allows for calculating the intersection accurately, with privacy consideration and without a third party. A simple protocol would go as follows:[3] once the parties agree on how to identify a person (let's say using the social security number), this number is hashed and shared with the other parties. If the other parties also have this hashed number, they will know that this member is shared and immediately know the identify. Remember that hashing is a one-way function, so if the hash is not found in the dataset of identifiers with their hashed value, it is nearly impossible to know what the identifier would have been.

The very first commercial application of SMPC reportedly stems from Denmark, where it was used in a sugar beets auction [58]. A real-life application that is similar to our salary calculation was performed in Boston in 2017 [55]. In a large fairness study, the difference in salaries across different categories,

---

[3] For more advanced protocols, have a look at the book by Evans et al. on SMPC and the references therein [136].

such as gender, ethnic background an level of employment, was investigated. Although this effort was supported by over 100 employers in the city, companies understandably would not want to make the salaries of their employees across these categories public. The SMPC protocol discussed earlier revealed the average salary per category combination, while not revealing the salary of the employees to any other companies or even a third party. Even though SMPC is a relatively young research domain, with current research focusing on making the technology more cost-effective [136], the different applications show the merits of these protocols for data science ethics whenever different parties wish to perform calculations on shared personal data.

4.1.5 Federated Learning

Whereas SMPC enables m parties to do joint analyses on their secret data without a third party, federated learning does include a third party, yet aims once more to preserve the privacy of the players' data. Federated learning is a distributed machine learning approach that aims to train a high-quality centralized model, while the training data remains distributed over a large number of clients [245, 295, 230]. How can this be? The short answer: by sharing the data science models, and not the data.

Consider the previously put forward case of Discussion 2, where a mobile OS wishes to conduct some analysis on data from a large set of mobile phones. Previously, we looked at the context of simply counting the number of times a certain emoji occurs, with the use of differential privacy. Now consider the case where the mobile OS wants to predict the probability of the emoji appearing given some words in a text, so as to be able to recommend the emoji in messages. With federated learning the mobile phones form a federation of players, who once again all would benefit from the shared data science exercise as the user experience would be improved, but also do not wish to share their private text messages with one another or with the third party.

The general setup is illustrated in Figure 4.4. It starts with a global model that is being sent to each of the devices (1). This model can be built on already available historical or public data, or can be a random model. Based on the local training data on each device, (2) a new local model is built on top of this. Next, (3) each device sends its local model to the third party, where (4) the previous global model is updated with the received local updates. Once applied, the local updates are removed as they are no longer needed.

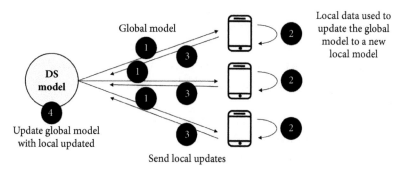

Fig. 4.4 Federated learning: a federation of devices update the global model with their local data. The locally learned patterns are sent back to the central server, not the local data. By combining the local models a new global model is created.

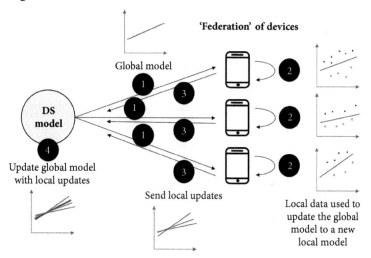

Fig. 4.5 Federated learning with a linear model and federated averaging.

A further refinement of this illustration is given in Figure 4.5 where we consider a linear model. Using the aforementioned case again, consider the learning of a linear model that predicts the probability for a certain emoji to occur, as a function of the words that appear in a text. On each device the linear model is updated based on the data on that phone. In the new global model, all the locally updated models are combined into a new global model (for example through an averaging of the coefficients, potentially weighted by the number of instances that each device used [295]).

Federated learning is especially useful to train models on data from mobile devices, or when the data is sensitive or very large (compared to the model). McMahan et al. propose a *Federated Averaging* algorithm to train deep learning networks [295]. The algorithm combines local stochastic gradient descent to update a model, performed by the device on the data of each device, with simple weight averaging to aggregate the local models into a new global one by the third party.

Even though the sensitive data now never leaves the device, there are still privacy risks. When for example, the device only has one text message, observing for which features (words) the weights have changed reveals the words of the one text message, which will likely allow to reconstruct the text message. The combination with SMPC or differential privacy could improve the privacy even further [295].

4.1.6 Summary of Privacy-Preserving Methods in Modelling Stage

We've now covered five different methodologies that enable the use of certain data science methods, while preserving privacy as much as possible in the modelling stage. These are summarized in Figure 4.6. They all attempt to reconcile the analysis of secret data with privacy. Key differences between these methodologies are the setup and the number of parties involved, the type of calculations that are envisioned, and related, the use cases. Only for zero-knowledge proofs and homomorphic encryption is the privacy perfect, but the practical use still involves difficulties. Once more we observe the trade-off between utility and privacy, though these techniques bring the two extremes closer to each other.

4.2 Discrimination-Aware Modelling

The use of algorithms and resulting prediction models make some people assume that the decisions are more accurate and consistent than decisions made by humans [270]. And while this can surely be true (at least if well-designed), it does not get rid of potential existing discrimination. When the data has a bias against certain sensitive groups, the resulting data science models likely will also have this bias. The use of such models in society then potentially lead to unfair decisions that can have a large negative impact on certain groups, for example when used to assess the risk of recidivism in a court ruling, or to

	Setup	# parties	Calculations	Use Cases
ε-Differential Privacy	■ ⊕	2	Aggregate statistics	Survey results
Zero Knowledge Proof	■ ⊕	2	Proof	Proof in range, Proof in set
Homomorphic Encryption	■ ⊕	2	PHE: + or × FHE: all	Cloud computing
SMPC	■ ■ ■ ■	m	Various (average, intersection)	Auctions, averages, voting
Federated Learning	■ ■ ⊕ ■ ■	m+1	Machine learning	Learning from mobile devices

■ Has a secret
⊕ Performs some calculations

Fig. 4.6 An overview of some privacy-enabling methods from the modelling stage.

assess credit risk in banking, or to decide who to recruit or admit in a HR or college admission setting. We'll start with definitions and associated metrics to measure the fairness of a prediction model, followed by methods to remove bias from a prediction model. Finally, some discussion cases are provided that describe the cautionary tales of using biased models in society.

4.2.1 Measuring Fairness of a Prediction Model

Once more we start with some metrics that allow to measure the extent of which a *prediction model* discriminates against sensitive groups. Previously the concepts of statistical parity and disparate impact were introduced, to measure the fairness of a *dataset*, for a dataset with positive outcome $Y = +$, sensitive attribute S where s is the sensitive value and ns the non-sensitive one. We now additionally have a prediction model M with predicted target variable Y'. The number of data instances from the sensitive group ($S = s$) that have a predicted positive outcome ($Y' = +$) is denoted as $s^{+'}$, and similarly the number of data instances from the non-sensitive group that have a predicted negative outcome

Table 4.1 Example dataset with 5 men ($S = ns$) and 5 women ($S = s$), with in total 6 positive outcomes ($Y = +$) of being hired. Three different classification models provide predicted labels Y'^1, Y'^2 and Y'^3

id	...	S	Y	Y'^1	Y'^2	Y'^3
1		ns	+	+	+	+
2		ns	+	+	−	−
3		ns	−	−	+	−
4		ns	−	−	−	−
5		ns	−	−	−	−
6		s	+	+	+	+
7		s	+	+	+	+
8		s	+	−	−	−
9		s	+	−	−	−
10		s	−	−	−	−

is written as ns^+. The total number of instances in the positive and negative group remains ns^T and s^T, respectively.

The metrics are similar in that they look at how the model's classifications are dependent on the sensitive attribute. They differ in what part of the dataset they consider to measure this. Let's revisit the example from before, predicting whether to hire someone or not, with gender as the sensitive attribute. Table 4.1 shows a dataset with five men ($S = ns$), five women ($S = s$), with Y indicating whether the person was actually qualified for the job or not, so the true labels. Several common fairness measures are introduced next [173], but be aware that many more exist [450, 152].

The four definitions of fairness, and related metrics (the difference between the ratios can be used as a metric) are demographic parity, equalized opportunity, equalized odds, and predictive rate parity, as given by the following equations. The fairness definitions will be illustrated for three classification models, predicting classes Y'^1, Y'^2 and Y'^3.

$$\text{Dem. parity}(M) : P(Y' = +|S = ns) = P(Y' = +|S = s)$$
$$\text{Eq. opp.}(M) : P(Y' = +|S = ns, Y = +) = P(Y' = +|S = s, \ Y = +)$$
$$\text{Eq. odds}(M) : P(Y' = +|S = ns, Y = y) = P(Y' = +|S = s, \ Y = y),$$
$$\text{with } y \in \{-, +\}$$
$$\text{Pred. parity}(M) : P(Y = +|Y' = +, \ S = ns) = P(Y = +|Y' = +, S = s).$$

A first common way to define fairness of a prediction model is *demographic parity* [189]. It requires that a positive outcome is independent of the sensitive attribute, which demands for our example that women and men are hired at the same rate. Or more generally: that the same positive rate is obtained for the sensitive and non-sensitive group. For model 1, with predictions Y^{r_1}, we have $P(Y^{r_1} = +|S = ns) = 2/5$, which equals to $P(Y^{r_1} = +|S = s) = 2/5$, and hence satisfies demographic parity. But as you notice, more women ($S = ns$) are qualified ($Y = +$) than men, yet the same percentage of women and men is hired. So if women were in general more qualified for the job, it is possible that fewer qualified women get hired (the same fraction as men who would get hired). It also does not look at how accurate the model is: even if persons are hired at random, the selection model can still satisfy demographic parity.

That's where *equal opportunity* comes in [189]. It demands that of the qualified persons, the same fraction of women and men should be hired. In data science terms: the true positive rate should be the same for the sensitive and non-sensitive group. So for model 1, we already noticed that the equalized opportunity is not satisfied, as the percentage of qualified women who are hired $P(Y^{r_1} = +|S = s, Y = +) = 2/4$ is not the same as the percentage of qualified men who are hired $P(Y^{r_1} = +|S = ns, Y = +) = 2/2$. If we look at the second model, with predicted classes Y^{r_2}, now equalized opportunity is satisfied, as the percentage of qualified women and qualified men hired is the same, being 50%. The name reflects that the opportunity to be hired should be the same for qualified women and men. Note that model 2 also satisfies demographic parity, as the probability to be hired for men and women is 50%. Yet, as is summarized in Table 4.2, the accuracy of model 2 is drastically less than that of model 1, already indicating that accuracy and fairness do not always go hand in hand.

A more stringent version of equalized opportunity is *equalized odds*, which not only looks at the qualified persons ($Y = +$), but also the non-qualified ones ($Y = -$) and demanding that the fraction of hired persons that are qualified and hired persons that are non-qualified is the same for men and women. So this definition not only considers the harm due to different true positive rates, but also looks at potential harm done due to different false positive rates. Neither model 1 nor model 2 satisfy equalized odds, but model 3 does. The

Table 4.2 The evaluation of the three models from Table 4.1 in terms of accuracy and fairness metrics.

Metric	Y^{r_1}	Y^{r_2}	Y^{r_3}
Accuracy	70%	50%	70%
Demographic parity	✓	✓	✗
Equalized opportunity	✗	✓	✓
Equalized odds	✗	✗	✓
Predictive parity	✓	✗	✓

fraction of hired qualified men (1/2) is the same as the fraction of hired qualified women (2/4), *and* the fraction of hired unqualified men (0/3) equals the fraction of hired unqualified women (0/1). Hardt et al. furthermore provide a method to adjust any prediction model so as to remove the discrimination in terms of equalized odds or equalized opportunity [189]. Roughly summarizing, the predicted labels can be altered as to ensure the fairness criterion, by using different cutoff values for different sensitive classes, while maintaining a good accuracy.

Predictive (rate) parity [90] is the fourth discussed measure, and requires the proportion of qualified men among all hired men to be the same as the proportion of qualified women among all hired women. In other words, given that they are hired, there is an equal probability for a positive outcome. Once more in data science terms: the precision rates of the sensitive and non-sensitive groups are the same. This is satisfied for model 1: of the hired men, both are qualified (2/2), and the same for the women (2/2). Model 2 does not comply, as only 1 hired man qualifies (1/2), while both hired women qualify (2/2). Model 3 agrees with the criterion (1/1 = 2/2).

A potential issue with both equalized opportunity and predictive rate parity is that they do not help to bridge the gap between the sensitive and non-sensitive group. Suppose that there are many more qualified men than women, with a limited number of positions to hire someone. To satisfy equalized opportunity or predictive rate parity, most hires will be men, thereby confirming the bias that exists in the data. The demographic parity strategy will hire the same number of men and women. Note that this same observation was previously described as a drawback of demographic parity, in the setting where there were more qualified women than qualified men. This finding illustrates that choosing the right metric depends on the context at hand.

The fairness evaluation shown in Table 4.2 of our simple example indicates that these measures can be conflicting. So, is it not possible to satisfy all of these fairness definitions? It has been shown that these can only be optimized at the same time under very strict and often unrealistic conditions, such as perfect predictability [242]. Wachter et al [455] argue that fairness measures that include the error rate (per group) implicitly favor the status quo, which is often not neutral and includes historical inequality. They therefore argue for measures that spur an open discussion on what discriminating factors are present and whether they are justified. Hence, it's important to be transparent about which measures are chosen and about the motivations behind these choices, while not judging others too harshly if they seem less fair for some groups according to some metric as this might be related to the chosen fairness definition.

Measuring the fairness of a prediction model requires having access to the prediction model, the sensitive attribute, but also a set of data points. So different datasets can lead to different fairness evaluations of a prediction model: a model could be fair when using the data of one location (or country), but not using the data of another location. So far we've looked at the fairness of a prediction model over a set of data instances, also called *group fairness*. We could also look at individual data instances, and determine whether a single prediction is fair, leading to *individual fairness* measures. Individual fairness is assessed by Dwork et al. as 'Any two individuals who are similar with respect to a particular task should be classified similarly' [121]. The risk is defining what similar means: if it misses important characteristics, such as certain grades or background, a new fairness issue is introduced. Counterfactual fairness [256] is another individual fairness measure. It looks at a counterfactual world for a certain individual: in another world where only the sensitive attribute changed (and causally related variables changed accordingly), the class would remain the same. The measure investigates the causal graph of variables and determines what the impact is of changing the sensitive attribute.

4.2.2 Removing Bias

There are several approaches to mitigate bias during model building. As briefly discussed next, this can be done by changing the objective function of the training algorithm, adding fairness constraints, combining different models, changing the thresholds used for final classification per subgroup, and approaches tailored towards learning a representation of the data that is invariant to the sensitive attribute.

If we want fairness, we should optimize for it. That is the idea behind the first set of discrimination-aware modelling methods. Kamiran et al. propose such a method for decision trees, which combines 'traditional' information gain in terms of the target variable, with information gain in terms of the sensitive attribute [234]. The final tree is obtained by additionally relabelling the leaves so as to optimize both accuracy and fairness. Another method is to include a regularization term in the objective function that penalizes discrimination, instead of, or on top of, the traditional regularization that penalizes complexity [235]. The objective function to minimize then becomes:

$$Loss = \text{training error} + \lambda_1 \cdot \text{complexity term} + \lambda_2 \cdot \text{fairness term.} \quad (4.4)$$

Kamishima et al. specifically apply this idea to a logistic regression formulation [235].

Zafar et al. [477] introduce fairness constraints that can be solved efficiently, and provide details on how this can be included in logistic regression and support vector machines. Fairness can also be obtained by combining different classifiers [76, 220]. An example ensemble method is Adafair, an extension of AdaBoost, which also includes fairness in the updating of the weights of data instances in each boosting round [220]. Yet another approach is to have the training algorithm learn a representation of the data that works well to predict the target variable, but not to predict the sensitive attribute [479, 480].

Kamiran et al. evaluate some of these approaches empirically and find that including fairness in the modelling stage of a decision tree leads to better results compared to the 'Massaging' method (described in Section 3.5.2) that works in the preprocessing stage. The preprocessing methods are generic, and allow any prediction technique to be applied, yet this decoupling can lead to a reduced accuracy, as compared to including fairness in the modelling phase. Most of the discussed discrimination-aware modelling algorithms provide a hyperparameter that balances between simply optimizing for accuracy, and the additional fairness optimization, which illustrates already the trade-off to be made between accuracy and fairness [152], or more generally: between ethical concerns and utility for the data. In a comparison study, Friedler et al. find that many of the evaluated fairness metrics are correlated, and no included method dominates all others on both accuracy and fairness [152].

An interesting post-processing case is made by Kleinberg et al. [241], similar to Hardt et al. [189], who find that rather than excluding sensitive attributes from the data, it is better to include them so as to have prediction model that is as accurate as possible, and then use different cutoff values per group (for the sensitive attribute) to include fairness. They demonstrate their theorem with a dataset on predicting college success to determine college admission. The data itself has a clear historical race bias, where college graduation rates are higher for white than for black students. By including race as a predictor, the most accurate ranking of individuals is achieved. Next, the race variable is used to determine different college admission rates for black and white people. By changing the cutoff (and hence admission rate) per group, admission levels per group can be made fair (for example by admitting the same percentage of black persons as there are in the general population). This idea makes sense: by including race, one is able to pick up on the different patterns that might exist for black and white groups that lead to good academic results. Secondly, it allows us to tune the cutoff so as to make the final results more fair.

It's worth emphasizing again that the included methods require the availability of the sensitive attribute, which might be difficult. Consider your bank asking about your ethnic background, so that they can make sure not to discriminate against your ethnic background. That might be a hard sell. In several cases, such as credit scoring, sensitive attributes such as race and gender are legally even forbidden to be used [264], making this solution very difficult to implement. Another challenge of this approach is upon explaining individual decisions. As the sensitive attribute has been included, it might well appear in the explanation for the decision (see Section 4.4), which of course is hard to defend towards a lay user.

4.3 Cautionary Tale: Predicting Recidivism and Redlining

4.3.1 Recidivism Prediction

In the US, predictive modelling systems are reportedly used that assess the risk of re-offending [13]. In one setup, the score is derived from a set of 100+ questions, either answered by the defendants or retrieved from their criminal records [13]. It includes questions as 'Was one of your parents ever sent to jail or prison?' or 'How often did you get in fights while at school?' but no question on race. The reason for using such prediction models is obvious: trying to come to consistent ruling and removing human bias from the system, by having 'objective' scoring systems determine a risk assessment based on patterns observed historical data. One could argue that this would reduce the rate of incarceration and allow the limited resources for treatment programs to be allocated to those who would benefit the most. Yet, dangers arise when such models are blindly deployed. A ProPublica article reports on a 2013 case involving a man named Zilly, convicted of stealing a lawnmower and some tools [13]. The prosecutor and defendant agreed on a plea deal of one year in county jail and follow-up supervision. Yet, the judge had seen Zilly's predicted risk score, which was reportedly high risk for future violent crime and medium risk for recidivism in general. The judge reportedly stated: 'When I look at the risk assessment, it is about as bad as it could be.' [13]. What followed was an overturning of the plea deal and a two years state prison sentence, with an addition of three years of supervision. This is just one anecdote that shows the potential consequences of using prediction models. The 2016 article reports that judges receive such assessments during criminal sentencing in several states in the US, including Arizona, Colorado, Delaware, and Washington.

If these predictions were perfect, then such a use might be warranted. But any good data scientist knows they are not. The data that the model is based on can be biased itself. The prediction model is often black box (and reported

Table 4.3 Different false negative rates (predicting high risk, but did not re-offend) and false positive rates (predicting low risk, but did re-offend) for black versus white persons, as reported by ProPublica in 2016 [12].

	White	Black
False negative rate	23.5%	44.9%
False positive rate	47.7%	28.0%

in the article to be 'proprietary' and hence not detailed to judges). The assessment, mostly done in terms of accuracy (with all problems such a metric comes with) has the risk of not investigating the performance for different sensitive groups. Note that this is no easy problem to solve: often bias is simply present in historical data, explaining prediction models can be very difficult (cf. Section 4.4) and detecting bias without having access to sensitive attributes is very difficult. Yet, taking these aspects into consideration, and transparently reporting on how these are dealt with, will get you a very long way and will improve the fairness of your models.

The ProPublica article specifically assessed the fairness of the COMPAS prediction model, and claimed that there are serious racial issues with such models that are actively used in court [13]. These models assess the risk that defendants and convicts are likely to recommit an offence within two years after being arrested. It seemed fair at first sight: the high risk black and white defendants re-offended at the same time. Yet, if we look at the reported misclassification rates, see Table 4.3 [12], we see that the errors are highly biased: black defendants were twice as likely to be misclassified as likely to re-offend than white defendants, and vice-verca: white defendants were twice as likely to be misclassified as likely to not re-offend. So again, reporting these numbers transparently, motivating the choice for fairness measure, and actively thinking through the potential bias in these systems are an important aspect of data science ethics.

4.3.2 Redlining

A well-known example of the unfair treatment of minority groups in credit scoring is *redlining* [351, 210, 144, 300]. It is commonly known as a practice where mortgage lenders are not giving loans to persons because they live in a predominantly black (or other minority) neighbourhood, and not because of their actual creditworthiness. Redlining can lead to disinvestment in minority neighbourhoods, with potential negative effects on property values and crime rates for decades [300]. Such policies even go beyond using proxies of ethnicity, as the red-zoned regions are often explicitly based on ethnicity.

> **Redlining**: 'Redlining refers to lending (or insurance) discrimination that bases credit decisions on the location of a property at the exclusion of characteristics of the borrower or property.' [210]

Historical housing discrimination was quite common in the US in the 1920s and 1930s [351, 210]. In response to the Great Depression in the 1930s, the US government created the Home Owners' Loan Corporation (HOLC), so as to provide low-interest mortgages to homeowners who were in default or lost their homes. Its parent organization produced security maps of cities scoring residential areas with a grade from one to four. Areas with African Americans were reportedly given a fourth 'D' grade, and coloured red. Although some researchers argue that these HOLC maps caused redlining, Amy Hillier argues that this is likely not true, and that lenders were avoiding these red-coloured areas well before these maps were created [210].

In Figure 4.7, such a security map for downtown Manhattan in 1937 is shown. These maps have been made public for various major US cities in an on-line tool by the Mapping Inequality project [378]. Each area was accompanied by a description, including a section on the inhabitants. An area description for area D3, known as East Village, states: *c. Foreign-born families 53%; Polish, Russian and Italian predominating*, while for area D26 in Harlem, north of 100th street, it states: *d. Negros: Yes, 90%* [378].

The term redlining is said to have been coined by John McKinght in Chicago in the late 1960s [144], where lenders would make maps with red lines drawn around the neighbourhoods they thought were susceptible to racial change and would not serve [351, 210]. The Fair Housing Act of 1968 prohibits such discrimination in the lending or home insurance process [444].

In 1988, reporter Bill Dedman from *The Atlanta Journal-Constitution* showed that the practice persisted through the 1980s [110]: Atlanta's banks would often lend to lower income white consumers, and refuse middle- and upper-income black consumers. The reporter was awarded the Pulitzer prize for his investigative reporting in this discriminatory matter [367]. One of the illustrations was a striking resemblance between two maps of Atlanta, one

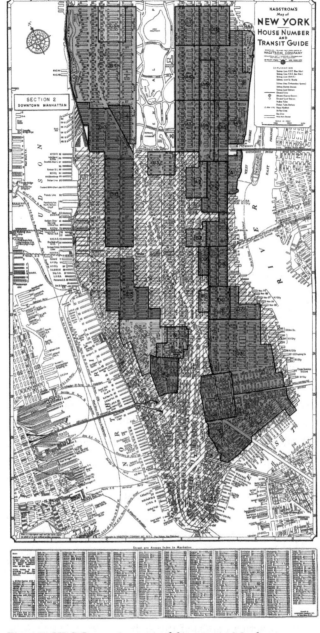

Fig. 4.7 HLOC security map of downtown Manhattan
(New York) in 1932, where the red zones were considered
'hazardous' for lending companies. Taken from the *Mapping
Inequality* project [378].

indicating where the black neighbourhoods are, and one where banks rarely lend [109].

So should banks never use location or neighbourhood at all? If it is not based on race, but rather on sound economic motivations, one can argue that the use is warranted. For example the extent to which the neighbourhood has access to public transport, the existence of flood zones or risk of earthquakes. Once more, there is a risk that these variables correlate with race. The previously mentioned fairness measures will make this transparent. For example, the reporting of Dedman shows the rejection rate of two banks in the 1980s, for both white and black persons, based on which we can deduce a demographic parity difference of 26% and 16% for the credit-scoring decision-making models of the two banks [110].

4.3.3 Summary of Discrimination-Aware Modelling

Two important assumptions are made here. Firstly, what constitutes a sensitive variable is given. Gender, for example, might be very interesting for targeted advertising, and hence included as input variable, for credit scoring gender may not be used. Similarly, race might be of interest in medical diagnosis, it is not used for credit scoring either. So make sure to know what your sensitive attributes are, and which you explicitly want to use. Remember that having access to a feature does not mean you should use it.

Secondly, to detect bias in your model, the sensitive attributes need to be present. If so, several fairness measures can be computed, such as demographic parity or equalized odds. These can be conflicting so one should think of the appropriateness of each and transparently report why.

The real danger is that data scientists build predictive models, but do not think about the potential biases against sensitive groups that are present in the model. By only assessing the overall accuracy of the model, and reporting good results, the risk is that users will blindly believe the model and assume it is objective and free of human biases. There is a need for transparency in the data used, the logic of the prediction model, what we consider to be potential sensitive groups, the predictive performance of the prediction model (including using misclassification costs and misclassification rates over the different sensitive groups), what is the most appropriate measure for fairness of the prediction model, and how the model is being applied.

The cautionary tales in predicting recidivism and credit scoring show how easily data science modelling can lead to blatant discriminatory practices.

Similar cautionary tales exist in image recognition, as already introduced in Section 3.4, where bias that is included in the data gathering stage, and not properly accounted for in the data preprocessing stage, will end up in the prediction model as well.

4.4 Comprehensible Models and Explainable AI

More and more attention in predictive modelling is shifting towards understanding how prediction models make their predictions. Consider for example tax fraud detection: tax administrations are increasingly relying on prediction models to assess the fraud risk of tax and customs declarations, with success [228, 447]. Zooming in on customs fraud detection at a harbour, the prediction model will flag a set of articles or containers to be looked at by a tax investigator (because of potential undervalued declarations, drugs, weapons, etc.). Understanding what exactly in the hundreds of variables has led to this prediction can be very helpful to know what to focus the attention to in the audit. Additionally, without an understanding of the predictions, investigators might feel puppets of the computer's 'black box' magic, simply being sent around by this artificial intelligence who unavoidably will make mistakes. A study on corporate residence fraud by Junqué de Fortuny et al. [228] looked to find which companies deceitfully attempt to place their residency in a low-tax country in order to avoid paying the higher taxes of their real location. A wide variety of data was used, from structured data, such as financial and location data, to fine-grained transactional invoicing data on which companies transacted with one another. The latter turned out to be very helpful: consider the following (fictitious) example when trying to predict which foreign companies likely have their actual headquarters in Belgium, but declare it abroad. Let's say we see that a foreign company receives invoices from a golf club in Bruges (Belgium). This could be an indication that the company and its owner(s) likely reside in Belgium. If this is indeed so, a foreign company that receives invoices from this specific golf club makes for interesting suspects.[4] When this company is flagged to an investigator, it would be very useful for her to understand that this is the reason for the prediction. Otherwise it would have to go through the hundreds of transactions of the foreign company, looking for a needle in a haystack. Knowing why the prediction was made would also allow us to quickly

---

[4] In a supervised setting, each company is a feature. A linear model would provide a weight to each company that indicates the risk for tax fraud when transacting with it. So the golf club in Bruges would have a high positive weight in this fictitious example.

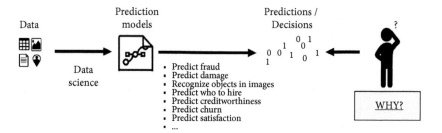

Fig. 4.8 Predictive modelling on large volumes of data can lead to accurate prediction models, as demonstrated in a wide variety of domains. Understanding how these prediction models come to predictions is of utmost importance in many domains.

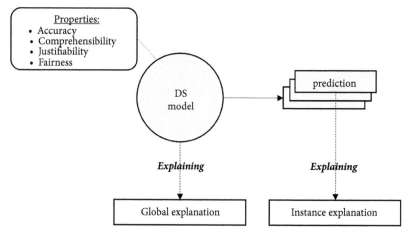

Fig. 4.9 Data science models have a certain level of comprehensibility, and lead to predictions and decisions being made. Explaining the model leads to global explanations, explaining an individual explanation leads to instance-based explanations.

overrule the prediction: if a typical sponsor of golf events, such as a manufacturer of golf bags, is flagged for this same business relationship with the golf club in Bruges, the investigator immediately understands why the prediction was made but also knows that this is a typical sponsor and hence can overrule the prediction.

This need to understand predictions is omnipresent, and is illustrated in Figure 4.8 with some examples. In predictive maintenance, when predicting that a machine is likely to break down and should hence get maintenance, understanding why can help to get insight into why machines break down, which technician to send, and what part of the machine to actually maintain. In image recognition, when a self-driving car predicts that there is a white cloud ahead, which turns out to be a white truck, it is important to understand why the model thought so, as to make sure this mistake is resolved. In HR analytics, when predicting which persons to invite for an interview, based on the words of the resume, it is important to understand why it is making such decisions, so as to ensure that there is no bias against certain groups such as women or immigrants. When predicting who to grant credit to, a bank needs to understand how these predictions are made, as they will need to explain these decisions to their applicants, cf. the opening story of this chapter.

Note that there is a distinction to be made between explaining to a user how the *world* works, so as to get insight into a certain domain, and explaining to a user how the *model* works. Although we focus on the latter, if the prediction model reflects reality with a high accuracy, then this would support understanding how the world works [287].

In the next subsections, we first need to clarify the difference between the terms comprehensible and explainable. Next, we'll go a bit deeper into why exactly we need to understand predictions, followed by a discussion on what makes a model comprehensible. Finally, a range of popular explanation algorithms are covered. So first things first: a short discussion on the nomenclature.

4.4.1 Understanding versus Explaining

Over the years, there have been many terms used in this area, with an absence of a clear consensus of the term to use: interpretable, comprehensible, transparent, explainable, intelligible, justifiable, intuitive, etc. (see e.g. [18] for some argumentation for, and definitions of, these terms). An important distinction to be made is that whereas comprehensibility (which we see as equal to interpretability and understandability in this book) is passive, referring to the property of a model that is often related to its complexity, explainability is active, referring to an action or a procedure that clarifies or details its internal function [18]. So a model is comprehensible to some degree, and can be explained by a method, as illustrated in Figure 4.9.

> **comprehensible**: *A property of a data science model (or explanation), which measures the 'mental fit' [277] of the model. Its main drivers are (1) the type of model and (2) the size of the model.*

explainable AI: *The research field that develops and applies algorithms to explain model predictions.*

More generally, a prediction model can have several properties. A first important one, which is often the only one considered, is generalization behaviour, or how accurate the predictions of the models are. Secondly is the comprehensibility of a model, indicating to what extent a user comprehends or understands the model. A related third property is justifiability or intuitiveness [284, 288]: to what extent is the model in line with existing domain knowledge. A model can be very accurate and comprehensible (for example, a rule predicting that people with large shoe sizes die earlier) but not in line with one's own knowledge. Yet another property is the fairness of the model, as introduced earlier, or the speed in which it can make predictions, the robustness of its prediction, and so on.

Figure 4.9 also reveals that there are two types of explanations: global explanations, where we explain a prediction model over the complete (or better a large proportion of a) dataset, or instance-based explanations, where we explain an individual prediction. The rationale for each is quite different. Consider a credit scoring application: the chief risk officer and regulator will want to have a global explanation of how the prediction model, which is about to get deployed, comes to *most* of its decisions. How such global explanations are obtained is discussed later, but consider for example a simple decision tree that makes the same predictions as the black box model in 95% of the cases: the tree can then be seen as a global explanation for the black box model. A customer on the other hand is likely not interested in why other customers were granted or rejected credit, or how the model globally works; she rather wants an explanation why she was just denied credit. Sometimes the model itself provides the explanation: for a decision tree, the path from the root node to the relevant leaf node will provide a rule describing why a certain prediction was given. For most other techniques, explaining requires additional steps, as described in Section 4.4.4. So keep in mind that an explanation algorithm requires as input a model and data: a global explanation algorithm will explain a model for a large set of data (mostly the training set), while an instance-based explanation algorithm explains how the model makes a prediction for a single data instance.

Now, what makes a model comprehensible? Defining comprehensibility is close to being a philosophical discussion. Comprehensibility measures the 'mental fit' [277] of the model, and its main drivers are (1) the type of output and (2) the size of the output.

4.4.2 Quantifying Comprehensibility

The first main criterion for comprehensibility is the model output, which can be rule-based, tree-based, linear, non-linear, instance-based (e.g. k- nearest neighbour), and many others. Which of these rule types is the most comprehensible is largely domain-specific, as comprehensibility is a subjective matter, or put differently: comprehensibility is in the eye of the beholder.

Ryszard Michalski is one of the first to address the comprehensibility issue in data science [299]. In his comprehensibility postulate, he states that 'The results of computer induction should be symbolic descriptions of given entities, semantically and structurally similar to those a human expert might produce observing the same entities.' Mainon and Rokach address this subjectivity issue as follows [277]: 'The comprehensibility criterion (also known as interpretability) refers to how well humans grasp the classifier induced. While the generalization error measures how the classifier fits the data, comprehensibility measures the "Mental fit" of that classifier.' This concept of mental fit is very convincing, and points out that if the user is more familiar with linear models, the mental fit with such models will be greater than the fit with tree-based classifiers. However, generally speaking, one can argue that the models that are more linguistic, will give a better mental fit. From that point of view, we can say that rule- and tree-based classifiers are considered the most comprehensible, and non-linear classifiers the least comprehensible.

For a given rule output, the comprehensibility decreases with the size [11]. Domingos motivates this with Occam's razor, interpreting this principle as [115]: 'preferring simpler models over more complex'. Speaking in a rule-based context, the more conditions (terms), the harder to understand [217]. There is a small nuance though: some might argue that a set of two rules is more comprehensible than one rule, as the former provides more information. Psychological research has shown that a human can only keep up to seven concepts in their mind at the same time [302], so from that size on comprehensibility decreases. This size concept can of course be extended beyond rule outputs [115]: the number of nodes in a decision tree, the number of weights in a neural network, the number of support vectors in a support vector machine, etc. Finally, though we consider model output and model size to be the main components determining comprehensibility, the concept can be deepened even further, as it also depends on aspects such as the number of

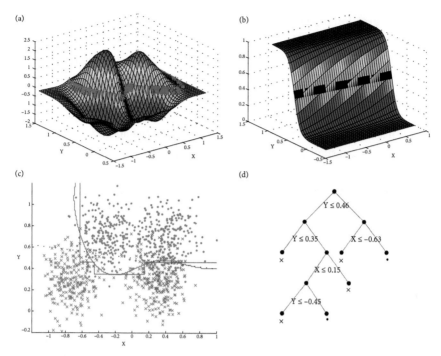

Fig. 4.10 (a) SVM and (b) logit prediction values, with corresponding two-dimensional (c) SVM (–) and logit (·) decision boundaries. Also shown is the decision boundary of a tree (–), with (d) the accompanying decision tree. Reprinted from Martens et al. [286], with permission from Elsevier.

(unique) variables and constants in a rule, the number of instances it covers, and even the consistency with existing domain knowledge [284].

Before continuing to methods to explain data science models, you might wonder: why are these complex prediction models being used in any case, given their large opacity issues? To demonstrate this often re-occurring trade-off between accuracy and complexity, have a look at Figure 4.10(c). This shows Ripley's data set, having two variables and two classes, where the classes are drawn from two normal distributions with a high degree of overlap [375]. Although this is a synthetic dataset, imagine it is a credit scoring dataset, with two input variables (for example normalized age and income) [286]. Each dot corresponds to a customer who took out a loan, the red crosses are customers that defaulted on their loan, the blue dots are good customers (did not default on their loan). The goal of a data science model is to build a prediction model that discriminates the blue dots from the red crosses. Let's consider three different techniques: a non-linear support vector machine (SVM) with RBF kernel, a logistic regression model and a simple tree. Figure 4.10 (a) and (b) respectively show the scores of the SVM and logit models. When we choose a cuff value of 0 and 0.5, this results in the red and black two-dimensional decision boundaries respectively of Figure 4.10 (c). It is already obvious that the red (SVM) line will be more accurate in its predictions as it is able to capture the non-linearities in the data. However, if we consider the output of the model—see Equation (4.5) where n is the number of training points, the αs are the Lagrangian multipliers, and σ is some constant, it is very hard, if not impossible to understand how the model comes to a decision: there are hundreds of terms where each term consists of an exponential function. Figure 4.10(a) also reveals that the impact of both input variables changes depending on the region of the input space. The tree that is built by decision tree algorithm C4.5 is shown in Figure 4.10(d), and is much more comprehensible.

$$f(\mathbf{x}) = \sum_{i=1}^{n} \alpha_i \, y_i \, e^{\frac{-||\mathbf{x}-\mathbf{x_i}||^2}{2\sigma^2}}. \tag{4.5}$$

Deep learning models similarly suffer from both a non-linear output and an even much larger size issue. The Google MobileNetV2 image net classifier, for example, uses a multilayer perceptron model with up to 6.9 million parameters [385]. However, such deep learning models tend to outperform all other methods in terms of prediction performance in domains as image and speech recognition [262]. So if these models are so accurate, why exactly do we still need explanations?

4.4.3 Why Do We Need to Understand and Explain Predictions?

The concept of wanting to understand something goes back to the beginning of humanity, and is the driving force behind scientific research. In the data science domain, explaining the workings of prediction models has been a research question for a long time. The importance of comprehensibility has been argued in the early 1990s by Kodratoff, who states in his comprehensibility postulate that 'Each time one of our favorite ML [Machine Learning] approaches has been applied in industry, each time the comprehensibility of the results, though ill-defined, has been a decisive factor of choice over an approach by pure statistical means, or by neural networks.' [243]. Today, this issue has been amplified, with the advance of deep learning techniques and the use of ever more complex data types [287, 307, 370, 451].

For businesses, our previous examples show that there are ample use cases for understanding the reasons why data science models make particular predictions or decisions. Prior work has suggested that when users do not understand the workings of the classification model, they will be sceptical and reluctant to use the model, even if the model is known to improve decision performance [239, 287]. In total, we can distinguish three reasons: (1) trust, (2) insights, and (3) to improve the prediction model.

A data science model that never makes a mistake is almost impossible. Any prediction model is likely to make some wrong decisions, but that doesn't imply such models shouldn't be deployed. Yet, in most data science applications, the predictions lead to decisions that come with quite an impact: on persons, on machines, on the bottom line of a company. And so we would want to be sure that we can **trust** the model.

---

trust: *Firm belief in the reliability, truth, or ability of someone or something.*
Oxford English Dictionary[5]

Trusting a data science models implies we believe in its reliability or truth. The (out-of-sample) predictive performance already indicates how accurate the predictions will be. Although we acknowledge the existence of mistakes, we want to make sure that the model has not learnt spurious patterns, or has learnt something that would only work in very specific circumstances. The explanation needs to provide further help in believing that the data science algorithm has learnt the 'true' pattern. It needs to provide an accurate proxy of the (complex) decision model while at the same time being comprehensible to the person getting the explanation [184]. Once this trust has been established, the model can be deployed.

A specific case of the need for trust is compliance. Due to the sensitive nature of certain applications, the ability to explain why certain decisions are being made becomes a regulatory requirement. Article 14.2.(g) of Europe's GDPR even states that data subjects have the right to obtain meaningful information about the logic involved when there is automated decision making [335]. The advisory organ of the EU on GDPR, Working Party 29, provided additional details on this concept of 'logic involved', pointing to the need for explanations. They write: 'The controller should find simple ways to tell the data subject about the rationale behind, or the criteria relied on in reaching the decision.

---

[5] https://en.oxforddictionaries.com/definition/trust

The GDPR requires the controller to provide meaningful information about the logic involved, ... The information provided should be sufficiently comprehensive for the data subject to understand the reasons for the decision.' [23]. The European 2019 guidelines on ethics in AI even state: 'Another great challenge is to clarify how to implement the requirement of explainability in a context where the complexity of AI algorithms can make it difficult to provide a clear explanation and justification for a decision made by a machine (i.e. black box effect).' [275]

The second goal of explanations is to obtain **insights** in a domain or into the model. This reason is part of one of the broader goals of predictive modelling: to learn something from a domain. For example, what is happening in my business: why are my customers churning? What are the main drivers of tax fraud? What are the main reasons we are rejecting credit to loan applicants? Such insights can also be useful for academic theory building, where data science models can lad to novel hypothesis and metrics [395]. Facebook like data has been used in such a manner. Kosinski et al. showed that what pages you like on Facebook is very predictive for various personality characteristics, with some insight on which pages are the most predictive [246]. To predict IQ, pages about humour, such as as Colbert Report, and the page Science are very predictive for a high IQ, while the same Science page is also very predictive for being dissatisfied in life. Praet et al. built further on this work, and predicted political preference using a Belgian sample [356]. They found that pages related to techno music (such as Dimitri Vegas & Like Mike and Hardwell) are predictive for right-wing political preference, while pages related to alternative rock music (such as Tom Waits or Fleet Foxes) are predictive for a left preference.

The third motivation for explanations is to **improve** the predictive performance of model. When working with complex models and data, detecting where data quality or bias issues are coming from is difficult. The gorilla problem, introduced in Chapter 2, showed that the problem is not easy to fix. An explanation of the prediction could reveal what exactly in the image led to the disastrous prediction: is it the composition, the white background (as potentially all gorilla images in the training set were these type of selfies with a white background)? Is it simply the presence of a face (if there were no images of humans in the training set, the algorithm would likely learn that any image with a face would be closest to the gorilla prediction)? Is it the lack of a selfie category? And so on. Such mistakes are far from unique. In a landmark *Nature* publication, researchers revealed how their deep learning network was capable of diagnosing skin cancer, matching the accuracy of 21 dermatologists and demonstrating the promise of AI in radiology [129]. In a following paper, the authors noted that images which included a ruler in the picture were more likely to be predicted to be malignant [317], as there were more images of malignant skin lesions with a ruler than for images of benign skin lesions. This bias might be explained as dermatologists often add this to measure the size when they are particularly concerned about the lesion [462]. An explanation for a malignant prediction would indicate what in the image led to this prediction—for example the ruler. This would provide guidance for data scientists on how to improve their training data, or, if the explanation follows the expert's opinion, provide trust that the model has learned the 'true' pattern. Also remember the story from the start of this chapter, where the decision on the assigned credit limit could not be explained. Another often described story is that of a military agency that trained an AI model to distinguish friendly tanks from enemy tanks [64]. Although the model obtained a high test accuracy, when deployed in the field, it had a very poor accuracy rate. Later it was discovered that the pictures of the friendly tanks used to train the model were taken on sunny days and the pictures of the enemy tanks on overcast days. So the system has simply learned to discriminate between sunny and overcast days. Gwern Branwen describes in detail the many versions of this story and tries to track down the origins. He concludes: 'AI folklore tells a story about a neural network trained to detect tanks which instead learned to detect time of day; investigating, this probably never happened.' [64][6] A similar story originates from the LIME technique (more on this popular explanation technique later), where the authors indicate that their deep learning model has learned to discriminate Huskies from Wolves, simply based on the presence of snow in the background. A bias that the authors intentionally included: 'We trained this bad classifier intentionally, to evaluate whether subjects are able to detect it.' [373] An explanation would indicate the snowy part of the image as the reason for the automated classification, thereby revealing the bias. A final example, by Vermeire and Martens [451]: using the pretrained Google MobileNet V2 image classifier, several images of lighthouses were wrongly predicted as missiles. One might imagine the shape of the lighthouse being the cause of this. Yet, the explanation reveals that the presence of clouds, in the shape of exhaust plumes that are often found behind a missile, is the cause of the prediction. Once more, this is a great insight for a data scientist to improve on the training set and resulting prediction model, for example by

---

[6] Although this story nicely illustrates the need for explanations, if the origin is unclear, should researchers and AI story tellers still use the tanks example of how things can go wrong? There are ample other, real-life examples, but if it is mentioned, it should be clearly noted that this is likely a fictional story (or find a reliable source).

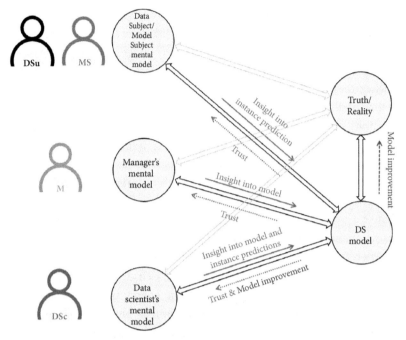

Fig. 4.11 Extended 7 model, by Martens and Provost (used with permission from *MIS Quarterly* [287]).

adding more images of lighthouses with such exhaust plume-shaped clouds in the background. To summarize, explanations can reveal for what data instances, and which feature (values), that mistakes are made, and can guide the data scientist on how to change the dataset so as to remove the mistake.

A conceptual summary of these drivers for explanations is provided in the 7-gaps framework (see Figure 4.11) proposed by Martens and Provost [287], which was inspired by the 3-gaps framework by Kayande et al. [239]. On the left side, the mental model of each of our user types is shown, while on the right we see the data science (DS) model, and the actual truth. Most often, the data science model will not fully correspond to reality, and wrong predictions are made. Improving the model can lead to a smaller gap between the data science model and the truth or reality. Obtaining insight into the model or into a single prediction can bring the users' mental model closer to the data science model. While the data science model can be moved closer to the users' mental model by providing trust and improving the model.[7]

---

[7] Three more gaps are proposed, which link the user types' mental models with the truth.

Table 4.4 Taxonomy of explanation approaches.

	Global	Instance
Feature Importance	Sensitivity Analysis	LIME/SHAP/LRP
Feature Value Plots	PDP	ICE
Rules	Rule extraction	Counterfactuals/Anchors

4.4.4 Explaining Prediction Models and Predictions

Explaining is active, ironically requiring yet more, complex data science algorithms to explain complex prediction models, all with the goal of coming to an interpretable explanation. A generic taxonomy of explanation algorithms is shown in Table 4.4. These algorithms differ according to (1) what data they explain, either a large dataset (usually the training set), leading to global explanations, or a single prediction, leading to instance-based explanations; and (2) the type of output they generate: the importance of the (main) features used by the model, a plot for each (important) feature that shows how the prediction score changes as the feature values change, and a simple rule or set of rules that describe the explanation. In what follows, some popular explanation algorithms are introduced. Note that further refinements of this taxonomy according to properties of the prediction model, data, explanation, or requirement can be made [237, 284, 307, 383, 409].

An important overall difference between the explanation approaches is what they explain: either the *decisions* (classifications) made by the black box, or the *prediction scores* of the black box. Explaining prediction scores will provide insight in how the prediction score changes in terms of the predictive features, while explaining decisions will provide insight on how certain feature combinations lead to a decision. This is an important difference to realize (as also motivated by Fernandez-Loria et al. in the context of instance-based explanations [142]). Data scientists have a tendency to focus on explaining prediction scores, as the final classifications depend on this prediction score, combined with a cutoff that can change with the context: have less marketing budget? Then target less of the likely interested consumers. Want a more conservative credit policy? Then make less credit granting decisions. This also motivates the widespread use of the AUC performance measure among data scientists to assess the predictive performance [364]. Model subjects on the other hand mostly are interested in the decisions that are made on them. If the decision corresponds to a negative outcome, they will very likely want to know what led to that decision, and how that decisions could change, rather than knowing how the prediction score could change, which might well or not lead to a different decision. From that point of view, data scientists can also be interested in explaining certain decisions: the misclassifications, as to improve the model. The rule-based approaches explain decisions, while the feature importance and plot-based explanation methods explain prediction scores.

4.4.5 Global Explanations

Sensitivity Analysis

An often used method is to simply look at what features are the most important for the model. It stems from the ease of interpretation of linear models, where the coefficient of each feature nicely describes the effect on the prediction score: one unit change in the feature value leads to a change in the prediction score that equals to the coefficient. But what if you don't have a linear model?

Let's briefly step back and see what that means. Return to Figure 4.10(a,c) and look at the non-linear SVM model. If we look at some fixed Y value (for example of 0.4), and we want to investigate the effect of the first X feature on the prediction score, we see that as we increase the X value from small values to larger ones, the prediction score will increase, but then decrease again, and later increase yet again. This is exactly the non-linear nature that makes it complex: there is no fixed, monotonic impact of the input variable on the prediction score. Phrased differently, the sensitivity or impact on the prediction score is not linear. One way to deal with this is looking at the feature contribution over a set of datapoints: if we change the value of the feature, what is the impact on the prediction score or accuracy *on average*? This effect might be different for different data instances, as these occupy different locations in the input space. Leo Breiman proposed this sensitivity analysis approach in his seminal paper on Random Forests [65], and it goes as follows: randomly change the values of a variable across the dataset and see what the effect is on the accuracy of the model. The larger this difference, the more important the variable [65]. A more advanced approach can perturb all subsets of input features and take the average marginal effect, inspired by the game theoretical Shapley values [454, 306]. Keep in mind that these feature importance methods indicate how sensitive or important a feature is in relation to the prediction score or accuracy and do not imply that the model is linear.

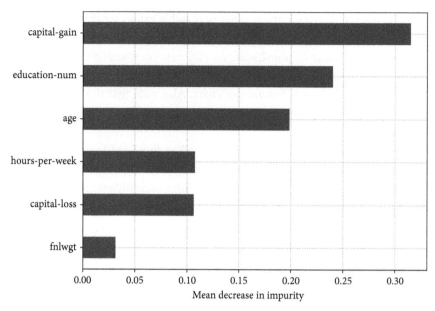

Fig. 4.12 Global feature importance for a Random Forest model, trained on the Adult Income dataset.

Another method specifically for random forests is the following: for each feature, we can look for all splits in the trees of the random forest in which the feature occurs, and calculate the decrease in impurity in each of these splits. The average over all these splits provides an indication of the global importance of the feature in the random forest. Figure 4.12 below shows the results for the Adult Income dataset, taken from the UCI dataset repository, which tries to predict if a person earns more or less than 50k per year, using six features, which are, in order of importance according to the mean decrease in impurity: capital-gain (income from other sources than salary), education-num (number of years of education), age, hours-per-week (number of hours worked in a week), capital-loss (spent income from investments) and fnlwgt (number of people belonging to the same group).

Plot-based
Plot-based methods rely on the power of visually interpreting a model [154]. A Partial Dependence Plot (PDP) is a two-dimensional plot that shows the marginal effect of a feature on the prediction score (vertical axis), over the complete range of the feature (horizontal axis) [307].[8] By showing the marginal average effect of a feature, PDPs take into account the interactions with all

---

[8] The plot can be extended to explaining the effect of two features using contour and colour maps.

other features, over the complete range of possible values for that feature. Such plots can neatly show the partial relationship between the input feature and the score, be it linear or non-linear. As PDP rely on visualizing data, human capacity constrains this approach to just two dimensions, thereby explaining only the effect of a single feature per chart. Also note that the chart remains an approximation, as PDPs assume that the explained feature is not correlated with the other features.

Figure 4.13 below shows the results for the Random Forest model trained on the Adult Income dataset, which shows that age plays an interesting non-linear role: it greatly affects the probability of having a high income if the person is around 40, and also has some effect for people around 60. Similar observations can be made for hours worked and capital gain.

Rule extraction

These methods extract rules from the trained black box model that are interpretable by humans and keep as much of the accuracy of the black box as possible, by mimicking the predictions. In its easiest form, the rule extraction method simply changes the class labels to the black box predicted labels, and applies a standard rule- or tree-based induction technique (such as C4.5, Ripper or CN2) on this relabelled dataset. So the result is a set of rules that mimics how the black box model makes its predictions. This approach has been shown to already provide good results, in terms of comprehensibility and fidelity [285]. The latter assesses how many of the decisions made by the black box model are explained by the rules, as measured by the percentage of test

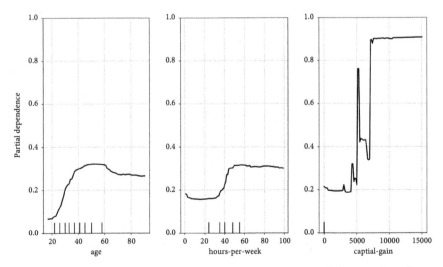

Fig. 4.13 Partial Dependency Plots for a Random Forest model, trained on the Adult Income dataset.

data that are given the same classification by both the extracted rules and the black box model.

More advanced rule extraction techniques usually build on top of this. Trepan is one of the pioneering techniques, proposed by Mark Craven in 1996 [97]. One of its key components is the creation of artificially generated data points at each split in the decision tree, which fulfill the constraints of the decision tree up to that split. For this set of data points, still a class label needs to be obtained. Trepan makes use of the black box as a labelling machine, which allows any number of data instances to be added. ALBA and ALPA are two techniques that build further on this idea, by adding data points close to the decision boundary of the black box model, as these are the input areas with the most prediction uncertainty [285, 105].

4.4.6 Instance Explanations

Feature value plots

The idea of PDPs to plot the partial effect of a feature on the prediction score can be extended to an instance level. For complex prediction models and heterogeneous data, it is possible that the relationship between a feature and the prediction score depends on the data instance at hand, which would ask for visualizing the effect per instance. Individual Conditional Expectation (ICE) plots address this by plotting one line per data instance, indicating how the instance's prediction score changes as the feature changes (keeping the values for the other features constant), whereas a PDP shows the average effect over all instances [168].

The example of Figure 4.14 (once more for the aforementioned random forest model) shows that the effect of age varies according to a person's characteristics. For example, people that already have a very high (and those that have a very low) probability of making more than 50K per year, don't see a big impact of age.

Feature importance ranking

The feature importance approach can also be used on an instance level: which features are most important for a specific data instance (located in a specific region of the input space). By focusing on a specific point, the problem becomes more straightforward, as we only need to look at the effect of the complex decision boundary in that location. LIME (Local Interpretable Model-agnostic Explanations) [373] does exactly that by generating additional training data around the instance to be explained, having a black box model score these, and subsequently build a linear regression model on top of this newly created dataset. (You might well notice the similarity to rule extraction, which also builds a surrogate model that explains the complex model, and generates

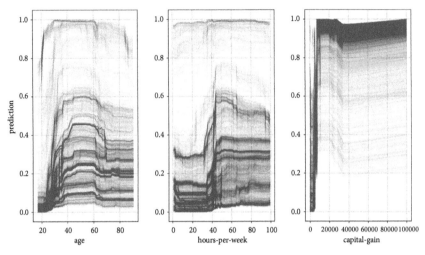

Fig. 4.14 Individual Conditional Expectation Plots for various instances' predictions for a Random Forest model, trained on the Adult Income dataset.

artificial data to be labelled by the black box.) The features with the highest (absolute) coefficients are then shown as the features most important for the prediction of the instance at hand. How many features are included is a user-defined hyperparameter. SHAP takes a similar approach and additionally ensures that the coefficients shown correspond to game-theoretical Shapley values [273].

Figure 4.15 shows that for a given instance with rather low probability of having a low income (32%), the most important features for this low prediction score are capital gain and to a lesser extent, education. The hours worked per week and capital loss are having a positive effect on the probability of having a low income, while age is the fifth most important feature in terms of absolute value of the linear coefficient in the LIME approximation, and contributes negatively to the prediction score.

The evidence counterfactual

Notice again that the previous instance-based methods explain a prediction score, not a decision. Explaining the decision of a predictive model can be seen as a causal question: what was it about this particular case (described by the data) that led the system to make the decision? That is, what evidence is in the data without which the system would not have made the decision: the evidence counterfactual or EdC (pronounced as 'Ed See')? An EdC provides a minimum set of evidence present in the data instance to be explained such that removing that evidence would change the decision [287]. Some examples: explain why I am shown this republican ad on Facebook: *If you had not liked*

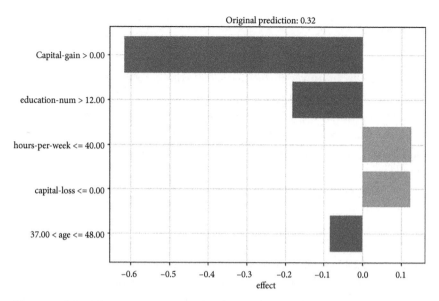

Fig. 4.15 LIME feature importance for the prediction of a Random Forest model on a certain instance of the Adult Income dataset.

the Facebook pages 'Fox News, Nascar and Trump' then the class would change from Republican to Center. Chen et al. argue that such explanations can help to decide on which of the Facebook likes should be 'cloaked' in order to inhibit the prediction [86]. An example from HR Analytics: when predicting which candidates to invite for an interview to a job opening, based on the words of the resume of the candidate, one would want to explain why someone is classified as being unsuitable for an engineering position. An example EdC could be: *If the words 'philosophy, McDonald's and COBOL' were not to appear on your resume then the class would change to suitable for an engineering position.* This provides insight that the model has learnt not to invite philosophy graduates that are (former) McDonald's employees and are working with COBOL. So the counterfactual is the datapoint that led to a different classification (for example a resume with certain words removed), while the explanation is the difference between the data instance to be explained and the counterfactual, which is the EdC (for example the removed words in the resume).

Counterfactuals have been used in philosophy for a long time already [10], and were introduced in the predictive modelling domain by Martens and Provost in 2014 [287]. In the meantime, dozens of novel counterfactual creation algorithms have been proposed [237]. As it explains a decision, an EdC provides guidance into what would need to change in order to come to another outcome, while not disclosing the model to the end user [40].

Discussion 8

The dean of your university has learned of your excellent data science skills and asks you to set up a new project. She wants to predict which students will end up in 'good' positions after graduating. This prediction model can then be used to identify high-potential students and provide them with additional 'extra credit' courses and personal mentoring.

1. What would be different definitions for ending up in a 'good' position?
2. How might bias against foreign students or women enter into the dataset, and the model? How would you assess the fairness of the prediction model?
3. Would an explanation of a prediction be needed for the students? And if so, what kind of explanation (approach) would you prefer to provide to a student?
4. Suppose the dean wants to collaborate with other universities in order to improve the accuracy of the prediction model, by using the data of students across different universities. Which of the privacy-enabling methods of Section 4.1 would you deem most suited to do so?

4.5 Cautionary Tale: Explaining Webpage Classifications

Martens and Provost describe a real-life case study from the domain of on-line advertising: classifying webpages as containing adult content or not, so as to avoid to allow advertisers to choose not to have theirs ads being shown on those pages [287]. A model was trained on a historical dataset of 25,000+ webpages with a total of 73,000+ unique words (the features, as a bag-of words approach was taken). Global explanation methods fall short with such high-dimensional, sparse data: both a set of rules or a ranking of the most important features would require thousands of features to be included in order to explain most of the predictions. The authors give the example of the word 'porn' being listed only as the 700th most important feature, so one would need to go over too many features just to find this obvious one. The most important feature, measured by the size of the coefficient of a linear model, is 'welcome'. More on this explanation in a second. Note that when a global explanation is still needed, for example by the manager of the deploying company, an innovative approach for textual and behavioural data was proposed by Ramon et al. [370]. She argues that instead of using the original features in the global explanation, one could use metafeatures, which group individual features.

Instance-based explanations were generated by Martens and Provost, in the form of EdCs [287]. Blindly trusting the reported 96% test accuracy could miss important insights and avenues for improvements. A couple of examples from their study showcase the three use cases. First, trust: most explanations of adult content webpages are quite intuitive, for example, *If the words 'searches nude domain adult' are removed then the class changes from adult to non-adult content*. Seeing this will reinforce the belief that the model has learned the true pattern. Some explanations provided insight in the situation: *If the words 'welcome fiction erotic enter bdsm adult' are removed, then the class changes from adult to non-adult content*. Upon closer investigation, this provides the insight that the first page of an adult content website often included a phrase in the line of 'Welcome, by continuing you are confirming you are an adult.' Yet other explanations reveal why the model is making mistakes: *If the words 'welcome searches jpg investments index fund domain' are removed then the class changes from adult to non-adult-content*. The algorithm likely has learned that a webpage that indexes several domains, contains links to jpg files, and includes a search function, will be an adult content page. Similar non-adult pages that index investment funds would therefore be wrongly classified as non-adult pages. Explaining these misclassifications can reveal such bias, which in this case would advocate having more of such non-adult content indexing websites in the training set [287].

4.5.1 Summary of Comprehensible Models and Explainable AI

Explaining predictions and prediction models are of great value. They can bring trust, insight into the domain and the model, and can lead to improvements in the data science model itself. Don't see the explanation as a necessary evil in a data science process, needed to ensure that there are no legal or reputational damages, but rather see it as making the data-driven decision-making process more transparent. Explanations can make challenges related to data quality and model bias visible, and potentially generate insights into the domain that reveal opportunities.

The explanations can be either global, explaining the model over a large set of data, or instance-based, explaining only a single data instance. The provided taxonomy further categorizes the potential explanation algorithms. Choosing which technique to use will mostly depend on the person that the explanation is for, and the context in which it needs to be given. Counterfactuals deserves a special mention, as they explain a decision made by a prediction model. Whenever an individual experiences a bad decision, be it credit being denied, being selected for a tax audit, not being hired or being convicted, the evidence counterfactual will tell why this is so. The intuitiveness of this explanation can then either comfort the model subject, or can provide grounds to object to the decision.

Other methods that explain prediction scores of course also have merit, definitely in any context where the threshold that is applied to the prediction score to come to a decision changes over time. Now you might be asking: do explanations of a prediction score not also explain a decision? Well, no: it has been shown that being an 'important' feature is not necessary nor sufficient to be part of an EdC explanation [142]. So think carefully when choosing the explanation method, keeping in mind what the use of the explanations will be.

Explanations are fundamental in data science ethics, and even have a legal rationale. Yet, explanations bring their own ethical challenges as well [40]: which method to use, for example. Some methods will provide several explanations. In the credit scoring example, it is possible that two counterfactuals for the same prediction exist: 'if you were not a woman, your credit would be accepted', and 'if your yearly income were increased by 50,000 Euro, your credit would be accepted'. The first would reveal a bias against women, the second not. A bank now has the power to ignore or obfuscate explanations that include the sensitive attribute, which in itself would likely be seen as an unethical practice.

4.6 Including Ethical Preferences: Self-Driving Cars

In the future we might want to add our own moral code to the decision-making model of machines. In this section we look at self-driving cars, the ethical dilemmas that these might face, and whether we wish to include generally observed human preferences.

A 2018 study at MIT studied the human preferences when dealing with the trolley problem [28]: if a driver could not break and is about to crash, should the car swirl to avoid a concrete block and kill three children, or crash the car at the expense of the driver's life? This study is inspired by the well-known trolley problem, credited to Judith Jarvis Thomson [124, 431]. The thought experiment goes like this: suppose a train trolley has lost its brake. Further down the track five persons are tied to the track and unable to move. You are standing next to a lever which controls a switch of the rails. If you pull the lever, the train will switch to a different set of tracks, but on that side track one person is tied up. You have two choices: (1) pull the lever, thereby saving the five persons on the main track, at the expense of the life of the person tied up on the side track, or (2) do nothing, thereby condemning the five persons on the main track to be killed by the trolley. What is the right thing to do? A utilitarian view of ethics would suggest that one is less than five, so pull the lever. A deontological point of view, however, would state that pulling the lever would make you participate in a morally wrong thing, and hence it would be better to do nothing. When this question is put forth to students,[9] most choose the utilitarian point of view. Yet, when they are confronted with the similar solution of killing a man to harvest his organs in order to save five other persons, doubts arise. Not surprisingly, this thought experiment, and many variants thereof (requiring to push someone off a bridge in order to stop the trolley, considering that the one person on the side track is your child, and so on) are the topic of many discussions in moral psychology and popular books [124].

The link with the MIT Moral Machine study is quickly made: provided with a dilemma of an imminent crash caused by failing brakes, who should the car save? Edmond Awad and his fellow researchers considered different types of subjects in the thought experiment, such as babies, children, cats, dogs, elderly, executives, and homeless people [28]. By putting such dilemmas forward to millions of persons from over 200 countries, a ranking of global ethical preferences could be established. The study found that, in terms of sparing, children were preferred over adults, adults over elderly people, and interestingly dogs over criminals. Important regional differences were also found in this study: the preference to spare the young over the elderly was much less pronounced in Eastern countries, such as Japan and Taiwan, as compared to Western countries, such as North America and Europe. Similarly did the authors find that Latin-American countries have a weaker preference to spare humans over pets. This highlights regional differences related to respect for the elderly or animals.

Then come the ethical questions on the use of these findings. Do we want these ethical preferences to be included as software modules in our future cars? And if so, do we want different modules in different parts of the world? Should there be legal standards dictating what the ethical preferences are, or should that be left up to the car manufacturers? Should we allow car manufacturers

---

[9] In the Data Science and Ethics class of the author of the book, in 2020 and 2021.

to put priority with the driver? And do we allow drivers to change the ethical preferences of their cars themselves?

There is some critique on the thought experiment, as its importance might be exaggerated and deflecting from other, more important ethical issues surrounding self-driving cars. A German Ethics Commission reflected on the ethics of automated driving [130] and stated that such dilemmas really depend on the specifics of the situation, and 'can thus not be clearly standardized, nor can they be programmed such that they are ethically unquestionable' [130]. The European Commission came up with a similar report in 2020, in which they said 'Interesting though they may be, moral dilemmas in crash avoidance are not the only, nor even the most urgent, ethical and societal issue raised by CAV [Connected and Automated Vehicles] safety' [212]. The interesting document recommends, among other points, to redress inequalities in vulnerabilities among road users, by for example having the car adapt (slowing down) to vulnerable road users, such as cyclists or young children. Other important ethical concerns that they mention when deploying self-driving cars include explainability, privacy fairness, and road safety. These additional concerns however still are well reflected in the call from the authors of the MIT Moral Machine [28]: 'we need to have a global conversation to express our preferences to the companies that will design moral algorithms, and to the policymakers that will regulate them'.

4.7 Summary

In this chapter we looked into the privacy, discrimination, and explainability aspects of modelling. We started with the different ways in which we can reconcile privacy with data science modelling. Differential privacy is revisited, now as a method to add noise to the analysis results, which leads to mathematical privacy guarantees, controlled by the parameter ε. Zero-knowledge proofs were shown to be a valuable method of proving some statement about a secret, without revealing the secret. Next we moved to homomorphic encryption, which might be an important component in future cloud computing. Finally, we discussed secure multi-party computation and federated learning as powerful methods to have many parties involved in privacy-friendly data science modelling. Keep in mind that having both 100% privacy and the full-fledged power of data science are hardly ever possible, requiring a trade-off that it tailored to your specific use case and privacy concerns.

In terms of discrimination-aware modelling, several metrics were introduced to measure fairness of a prediction model. Unfortunately, these often conflict with each other, making the choice of measurement another ethical decision that transparently needs to be reported. Enforcing fairness to some degree can be done by changing the training data, adding fairness constraints to the objective function, changing the thresholding to come to final decisions, or learning discrimination-aware representations.

The third part of this chapter looked at building explanations and comprehensible models. The irony of this work is that we want to understand predictions, by applying complex algorithms to complex prediction models. The taxonomy of explanations provides guidance on which approach to use, based on the setting and user type: managers often want global explanations, while model subjects are argued to favour instance-based explanations.

Let's now revisit the opening story of this chapter, where an apparent discrimination in Apple Card decision making was voiced [5, 318]. This outcry has two components. First, the automated decision requires an explanation, so that such statements as 'It's just the algorithm' or 'Computer says so' are simply never needed. This specific setting would have benefited from a counterfactual explanation: stating what evidence in the data from the man would need to have changed to lead to a lower credit limit decision (perhaps the high limit is due to a long-standing credit history or previous job history that the woman might not have had). Being able to immediately provide such an explanation, requiring mostly automated explanation algorithms, would have either eliminated the worries from the customer, or would have shown that indeed some 'wrong' decision was made due to bias. If the counterfactual had been: 'if the customer would have been a woman, then the credit line would be 10 times lower', the bias against women would immediately be apparent, warranting changing the decision, and the prediction model as a whole. The fairness of the model could also be investigated with the demographic parity measure: is the positive outcome (a high credit limit) independent of the gender of the customer? Having done this initial basic test would have allowed the company to respond swiftly that no such discrimination was observed in the model.

5

Ethical Evaluation

In his free time, Danny the data scientist has been analysing data from Reddit, a website with user-generated content, in an attempt to predict the stock market.[1] He's been reading about personality traits and investment profiles, and decides to create 500 models that each predict any combination of personality and investment sentiment, based on the text he crawls from the Reddit website. Danny believes these sentiment scores could correlate with, and predict, stock market movements. After much data crunching, he comes to his findings: one of the created scores has an accuracy in predicting the NASDAQ of 75%—jackpot! His work is picked up by the media, where Danny frequently provides interviews explaining his work, and a fund is even being created by an investment bank that trades according to his score. A year later, the fund no longer exists and Danny no longer talks about this work. What happened?

When it comes to ethical evaluation, three important questions arise.

1. First, what to measure? In our stock market prediction example, is accuracy the right metric for the stock prediction model? On what set was it evaluated? What benchmarks have been used? Do we need to evaluate any of the FAT criteria?
2. Second, how to interpret the results? Are the findings significant? Might p-hacking be happening, or did multiple comparisons happen without properly correcting for them?
3. And finally, what to report? Is the process completely transparently reported, both the good and bad results? Is the data science reproducible, and how easy is it? We'll go over each of them in detail, next.

5.1 Ethical Measurement

Ethical measurement involves two components, being a good data scientist, and being good. In the first part the emphasis is on *data science*, which should

---

[1] This is a fictitious story.

be done correctly. So being aware of the different evaluation techniques and metrics that should be applied to measure the performance of data science models. In the second part, the emphasis is on *good*, not cherry-picking or simply lying about data science evaluation metrics for merely one's own gain.

5.1.1 Correct Evaluation: Doing the Data Science Right

The power and danger of data science evaluation is that different measures can lead to very different findings and conclusions. We already discussed the importance of a proper definition of the target variable in Section 3.3.2. But how do we now evaluate how good the predictions actually are?

It starts with basic data science. As you've most likely seen in previous introductory courses to data science, one must always use a test set that is as representative as possible for the population on which the model will be applied. This set is preferred to be out-of-time (versus out-of-sample) and should be large enough. Not using a test set, or using a test set that is not representative, is just bad practice and can lead to an overly positive estimation of the impact of data science. Such bad practices are not necessarily sprung out of a conscious decision; perhaps it is even an oversight because of initial good results that excite you as a data scientist so much that you lose sight of the critical questions you're supposed to ask yourself on the evaluation. So being ethical demands you are a good *data scientist*.

Suppose we'd ask Danny if he used a test set, and how it was chosen. His answer reveals there was a test set, chosen out-of-time: he trained on data from January to November, and then used four weeks of December as a test set. His prediction model was correct in predicting whether the NASDAQ went up or down that day in 15 out of the 20 days. Hence the test accuracy of 15/20 = 75%. Immediately you see a first issue: December might not be representative for a whole year of trading. The potential seasonality of the stock market is reflected in the saying 'Sell in May and go away. Don't come back until November' [222]. But on top of that, Danny has evaluated his score on 20 days only. This seems a very limited time period, likely not large enough to demonstrate the predictive performance upon deployment.

Next to the chosen test set to evaluate the data science model on, there might be issues with the accuracy metric itself. This metric is very intuitive for business and end users alike and is therefore (understandably) often used to communicate the results to a broad public. But it has several problems, such as class imbalance and not taking into account the different misclassification costs between false positives and false negatives. If in Danny's test set of 20 days, the NASDAQ increased on 16 days, a simple model that always predicts the market will go up would have an accuracy of 16/20 = 80%, even outperforming Danny's model. But would you want to trade on this model? Of course not, as it would recommend to buy the NASDAQ every single day, forever…

Other, more suitable metrics and graphs include the confusion matrix, the ROC curve, lift or profit curves, and summarizing metrics thereof [364]. The performance also depends on the chosen threshold to make a positive decision. So even if ROC curves are used, one should still be able to motivate the final chosen threshold used, to go from a prediction score to a classification. Additionally, consider several baselines models such as a random model, a majority vote model, an expert model, or whatever seems logical for your application. A suitable baseline for Danny could have been a model that as a prediction gives whatever the market did the day before: if the market went up, the prediction is it will continue to go up, if it went down, the prediction is that the market will go down the day after. Even if the target audience of your report consists of lay users, who have no idea what an AUC is (nor have an interest in being lectured on this), be sure to have done the proper analysis yourself and be able to motivate to fellow data scientists why the chosen model is the right one.

These choices are mostly just good practices of data science, but also have an important ethical component if these decisions would be taken consciously. Beating existing models and baselines can be very hard, so there is a risk that only cherry-picked metrics are reported that provide a good story, knowing that upon deployment this probably won't hold. As a data scientist it is too easy to choose an easy-to-predict test set and report a single metric that suits you well. Definitely towards non-experts, such reporting would go unnoticed.

5.1.2 Evaluating FAT

Next to doing the *data science* right, one should also focus on the ethical, *right* part of the evaluation. We've covered the FAT principles of privacy, discrimination against sensitive groups, transparent process, and explainable AI, in the different stages discussed so far. Whenever there is personal, sensitive data, or sensitive groups in the dataset, consider the fairness measurements, the privacy assessments, and the comprehensibility requirements. Report transparently which sensitive groups were considered, what data you think might be personal and what possible re-identification methods would look like. Based on these you can assess the privacy of your dataset, in terms of k-anonymity, l-diversity, t-closeness, or the fairness towards sensitive groups in the dataset, using measures as statistical parity, or of the model, with metrics as demographic parity or equalized opportunity. The extent to which model subjects should be entitled to an explanation, and to what extent this explanation can be provided, should also be evaluated.

For some applications, like Danny's, who uses public Reddit data for stock market prediction, there might not be an issue related to data subjects or model subjects' privacy or discrimination. But what if Danny was using payment data from the bank where he is working? This is part of the thought experiment that is considered in the next discussion.

Discussion 9

You're the Chief Data Science Officer at a large bank. You've instructed your team to experiment with using payment data for marketing purposes, predicting which customer might be interested in a golf tournament that the bank sponsors. So the data instances correspond to customers, and the features are unique account numbers. Your newly hired team is ready to shine and has put quite some effort in building a linear model, where each account number that one can pay to is given a coefficient. The prediction model hence predicts interest based on whom the customer has made payments to. They proudly report to you that the accuracy of their model is 95%, on a test set chosen in January.

1. What further questions would you ask on the evaluation? Think of test data, metrics, and baselines.
2. What would be potential privacy risks related to re-identification or the revelation of sensitive information of customers to the data science team? How to measure these?
3. Might there be discrimination against sensitive groups, such as Muslims or women, if the payment data is used? How to evaluate? Might there be certain features (account numbers) that if a customer made a payment to those, the sensitive attribute is revealed? How to measure whether the model is using these in a discriminatory way?
4. Would the invitees of the golf tournament event require an explanation for their predicted interest? If so, what type of explanation would you provide?

5. How would your answers change if the target variable was now credit risk (defaulting on a loan or not) and the data is provided to an external academic research group?
6. Would you expect your data science team to have answered (or at least raised) all the previous questions, when they report their findings?

5.1.3 Evaluating Other Ethical Requirements

So far, we have focused on the FAT ethical requirements, which arguably encompass most of the ethical issues that exist in the current day and age. There are some additional ethical requirements that have been proposed, which are of importance in certain applications.

Robustness
In the European 'Ethics Guidelines for Trustworthy AI', robust AI is positioned next to ethical AI and lawful AI, the combination of which is said to lead to trustworthy AI [203]. On robustness, the document states: 'Technical robustness requires that AI systems be developed with a preventative approach to risks and in a manner such that they reliably behave as intended while minimising unintentional and unexpected harm, and preventing unacceptable harm.' [202] Robustness includes resiliency to attacks, fallback plans, accuracy, reliability, and reproducibility [202]. Several of these components are related to previously mentioned ethical evaluation measures, such as security, accuracy, and reproducibility. A fascinating related research topic is that of adversarial attacks and how to make model resistant to them [276]. Adversarial attacks will change the input (for example an image) so that the change is almost indistinguishable from natural data, yet the predicted output becomes something unexpected. Robustness of a model evaluates the resilience against such attacks.

Sustainable
The massive calculations of modern-day data science models, and then mainly deep learning models, come not only with an economical cost to train the models on large computers, but also an ecological one, as both the training

and deployment of models can require quite some energy. For those large number-crunching applications, data scientists should consider its effect on the planet's ecosystem and environment. Consider the well-known GPT-3 language processing model from OpenAI, for example, which can write creative fictions and make translations [70]. This deep learning model has been trained on 500 billion documents and consists of 175 billion parameters [70]. A 2019 study estimated that training a single deep learning model can generate as much pounds of CO2 emissions as the lifetime carbon footprint of five cars [416]. At that time, GPT-2 was the largest model available for the study, and 'only' had about 1.5 billion parameters [70].

The rise of deep learning models will therefore lead to questions as: can we do the same with less computations, or can we do something almost as good with less energy usage? How much less, and what is the ecological impact? These kinds of evaluation are of importance when training a system that is doing massive computations (like GPT-3), and/or when a data science system is being deployed across a very large number of machines. Think for example of the deployment of deep learning models on mobile devices. Investigating and optimizing the energy cost of such models, when deployed on smartphones, can potentially have a relatively large effect on aggregated energy usage and carbon emissions. The deployment of data science in an 'Internet of Things' or self-driving cars setting makes energy efficiency both an economical and ethical issue. Being sustainable doesn't require you not to use data science; it rather means that you should be aware of this issue, be transparent about the energy consequences (potentially with tools such as the *Machine Learning Emission Calculator* [257] or *carbontracker* [14]), and think about how to improve on it. For example, by limiting the search space of hyper-parameters to relevant ones, limiting the number of wasted experiments, choosing your data centre and cloud provider based on the carbon footprint of calculations, and using energy-efficient hardware [257].

5.2 Ethical Interpretation of the Results

To evaluate the significance of the results, statistical tests can be applied to determine whether the results are robust. A typical comparison is: is the result significantly better than the baseline model? The p-value then indicates whether this is the case or not, according to some chosen statistical test. There are two ethical considerations to be made when interpreting these results: p-hacking and the issue of multiple comparisons.

5.2.1 p-Hacking

A p-value is defined as the probability of obtaining results at least as extreme as the observed results of a statistical test, assuming that the null hypothesis is correct [24]. So it will assume some theoretical distribution of the measured performances and based on that determine the p-value. p-Hacking occurs when 'researchers collect or select data or statistical analyses until nonsignificant results become significant' [195]. The practices we'll discuss can also be used when tuning our data collection, preprocessing, modelling and evaluation on a test set, a practice which is sometimes also called data dredging.

One typically needs several observations to conduct a significance test, for example by repeating your experiments 10 times, each time with a different, randomly chosen test set. This brings forth a first issue: what if you notice that for one fold, you happen to have a much worse performance than for all others? You reason that although the folds have been chosen randomly, this specific test set is very difficult to predict (and messes up your good results), so you decide to do another random division of your dataset into 10 folds, and rerun your experiment on the new test sets. Now it looks better. But you don't stop there. Instead of changing all folds, you consider the test fold with the worst performance and change this with a new randomly chosen fold, hoping the performance becomes better. And repeating this process until all folds have a similar maximum possible performance. This is clearly unethical data science. The evaluation should be aimed at getting an independent evaluation on how well the model will work on new, unseen data points, not on how to squeeze out the best performance for your model by gaming the evaluation setup. This is already a first example of so-called p-hacking: messing with the performance assessment to get significant improvements.

Once we have our set of performance measurements (let's say we measure accuracy), the statistical test can be applied to test if the accuracy of our model is the same or not larger than the accuracy of the baseline model. A cutoff value α is typically chosen to be around 5%: if the p-value is lower, we can say that the difference is significant. In that case, the probability that we would see this if the results were *not* significantly better is less than 5%, so it is a robust and significant finding.

In data science, tuning your data or model to get the right evaluation and p-value on the *test* set is a big no-no. There are several ways to p-hack, of which we already introduced one (messing with the test set), depending on where in the data science flow additional intervention is done. The first is in the data collection and preprocessing stage, where one adds instances to the dataset[2] until the right p-value is obtained, discarding all other instances. This can even be done more cleverly (or deviously), by seeing which instances should be included in the dataset, so that the right p-value is obtained on the test set.

A second way to do p-hacking is at the input variable level: applying transformations to the variables and/or doing input selection as to get the right p-value. Let's keep in mind that input selection or transformations on the input variables are not necessarily a bad thing (*au contraire!*), as long as it is evaluated on a validation set. When it is done to get the best (significant) results on the test set, it becomes unethical p-hacking.

The third approach is in the evaluation itself, where many evaluation metrics are tried, and only the one that is significant is reported, no matter whether it is the right metric for your application. Danny might have tried to assess its model in terms of expected profit, lift, F-measure, and accuracy, and found that accuracy has the best results, so he reported only that one. One can furthermore test the significance of the test performance at different thresholds, and report the best one only. Yet another p-hack is creating many prediction models but only reporting the one that performed best (most significant/lowest p-value), an issue we will focus on in more detail next.

As you notice, there are plenty of ways to get significant results on the test set, so as to demonstrate significantly improved results, but these are just bad science and business practice. So why would people do p-hacking? It's clearly the wrong thing to do. In research, a negative finding, or a novel method that is not significantly better than what is already out there, is very hard to publish. Similarly in industry, a data scientist would of course rather report great results, with significant impact on the business, than reporting that he or she was unable to improve on the existing system. Having good results is simply what we all aim for, in any aspect of our lives. When you want an academic promotion, or get your work published in a good journal, or simply obtain more funding for your research or department, p-hacking is an easy yet unethical practice to get these, by changing bad findings into good ones. Even if there is no direct promotion or funding impact of the results, having much better performances can lead to fame and glory, which most of us are sensitive to as well. The fame will be short-lived when you are p-hacking, as the bad results are bound to be revealed upon deployment or when the experiments are reproduced by other persons.

---

[2] The dataset comprises of the training, validation, and test set.

So be aware of this practice, as to recognize the risk with others but also with yourself. If we don't actively combat *p*-hacking, there exists a potential disaster for both scientific and business progress, as we are no longer being led by actual results in our decision making.

5.2.2 Multiple Comparisons

Next to the active messing up of the evaluation process to get the right *p*-value, there is also the issue of doing many comparisons and interpreting results as significant, simply because some models have a *p*-value below the chosen α of 5%. The issue with multiple comparisons is that the more tests you conduct, the higher the probability that some rare event occurs by chance. Consider for example the experiment where you throw 1000 one Euro coins in the air. Upon investigation of the coins, you notice that some are cluttered and all heads. This will be due to chance, not because you have magical throwing abilities.

Let's revisit the *p*-value meaning: suppose your test finds significant results at $\alpha = 5\%$. Then what is the probability that this finding is due to chance, so there is no real improvement? Well, it's 5%. Now suppose that I build *m* prediction models, and test for each model whether the accuracy is significantly better than random guessing, at 5% level. What would the probability be that *at least one* model is significantly better than random guessing, simply due to chance[3]?

$$P(\text{at least one significant result}) \quad = \quad 1 - P(\text{no significant result})$$
$$= \quad 1 - (1 - 0.05)^m.$$

If we do the experiment just once ($m = 1$), the probability to have at least one significant result (due to chance) is 5%, which makes sense, as we are measuring significance at a 5% level. If we build 10 models, there is a 40% chance that at least one model will have a significant result, even though the results are not significantly better. With 20 models this already becomes 64%, while if we build 100 models, there is a 99% (!) chance that at least one of these models is significant even though the tests are actually not significantly better. Overseeing this issue has led to a realization that many of the statistically significant results that are reported in (academic) publications likely do not hold up [162].

So returning to the 500 models that Danny has built using Reddit data to predict whether the NASDAQ goes up or down. Danny proudly reported that

---

[3] Example inspired from the Statistics example of Goldman [166].

the results of one of these prediction scores is significantly better than using a baseline model of predicting for the next day what the NASDAQ did the day before, even at a 1% level. Given the previous calculations, this doesn't mean much, as we can be almost certain ($1 - 0.99^{500} = 99.34\%$) that at least one of the 500 models will be significantly better at a 1% level, just by chance.

The solution to correct for these multiple comparisons is the Bonferroni correction [345]. When m different hypothesis are being tested, instead of using α, previously chosen as 5%, use α/m as the cutoff to conclude upon significance. When 10 models are built, the significance is measured now at a 0.5% level instead of at 5%. If we then calculate what the probability would be that at least one of them is significant due to chance, we get 4.9% ($= 1 - 0.995^{10}$); very close to the 5% level we initially aimed for. The Bonferroni correction itself is not perfect, as it assumes independence of the individual tests (and prediction models in our case). This is not always the case and therefore can lead to a higher probability of false negative findings [345].

The overarching solution for both p-hacking and the issue of multiple comparisons is once more transparency: simply report transparently all the steps taken (which models were built, how, on what dataset, etc.). Such ethical reporting is discussed next.

5.3 Ethical Reporting

5.3.1 Reporting Transparently

Ethical reporting boils down to reporting transparently, both the good and the bad. Don't just reveal the success stories, document the complete data science flow, and motivate each step, including the ones that didn't work out. Such a report should answer such questions as:

- **data instance level**: why did you choose a certain sample size? Is the sample representative? Why do you think so? Did you look at learning curves (the impact on the performance as you add more data)?
- **input variable level**: What variables did you consider and why? How are they obtained? Did you remove variables? Why, and how?
- **modelling level**: What techniques did you apply? Were they tuned? What models did you test?
- **evaluation level**: What evaluation metrics were used and why? And so on.

This all relates to knowing, applying and reporting proper data science practices. The book by Provost and Fawcett is a great book to get the basics of data science in a business setting right [364].

The reason that this transparent reporting is so important is to ensure trust in the data science that went into building the model. Trust that the data scientist is knowledgeable and conducted the right data science practices. But also trust that the data scientists did not consciously game the system, with practices as p-hacking or tuning on a test set.

To make this more tacit, aiming for reproducibility can be of great help. If the reporting allows us to easily reproduce the findings, trust will naturally emerge, confirming that the data science actually works, makes sense, and can be built upon. Reproducibility involves having the data and code available and easily accessible, where a main script automatically goes through all the data science steps and leads to the same results as are reported. Within businesses, reproducibility also assures that the work is not lost when the data scientist leaves the company. Undocumented code, or code that does not lead to the results that were reported, will at least lead to efficiency losses, but can even lead to the negation of previous findings. If these earlier results have already been deployed, they would need to be discontinued, with all the financial and reputational losses that go with it. Any data science reader that had to read and continue on undocumented code of another data scientists will likely be able to confirm the frustrations that go with improper reporting.

Academics are also encouraged more and more to share the data and code when publishing their findings. Sharing sensitive or personal data (even if 'pseudonymized') should be done with much care, as discussed in Chapter 3. There are of course good reasons not to share the data, for example because the data was obtained through a collaboration with a private or governmental partner, or personal data was gathered during the research. Code sharing is encouraged through platforms as github, and even helps in the marketing of one's research findings.

An interesting tool to report on the key components of your data science model is a so-called 'model card'. This framework was introduced in a paper by Google employees in 2019 to encourage transparent model reporting [305]. Model cards tend to be one or two pages long and make clear what the intended use is of the model, what training data was used, the model type, how it was evaluated, limitations, who can be contacted for more information, and other important model properties.

Let's get back to Danny's stock prediction model. In his report, Danny should have motivated why he only chose 20 days of test data, in December. Is it because of a lack of data? But the data is publicly available, both the input (Reddit data) and target variable (NASDAQ) so that would probably not be the real reason. Is it because he just wants to use his model in December? But why would he? What arguments does he have for this being a representative sample to evaluate the model on? On the evaluation metric, why is the test metric the right metric to use, and did he try any others? In the trading community, risk-adjusted metrics are often used [226]; were these considered? Why (not)? Similar motivations on the data preprocessing and modelling should be provided. On the more general transparency requirement of reproducibility: is his data and code publicly available, so that other researchers can verify the findings, and even better, can apply the model and verify if it consistently performs well, beyond his chosen test set? Danny would very likely have trouble showing and motivating this, immediately revealing to himself that what he found was not actually correct, and it could guide him to further improve the data, model, and evaluation so as to properly demonstrate he actually can (or cannot) predict NASDAQ movements better than some chosen baseline.

5.3.2 Ethical Academic Reporting

The world of academic reporting is a special one and merits additional attention. The incentive structure and academic reporting processes can easily lead to 'small' unethical practices, such as cherry-picking the datasets to report on in academic publications, or even fabricating results altogether.

Reproducibility as a sign on the wall
In an academic study, a researcher will tend to posit its hypothesis or research question, gather and analyse data, and present the results. Readers of these publications will trust that the paper has been truthfully written, and that the same results could be obtained when going through the exact steps as written by the authors. Unfortunately, too many stories demonstrate that this is not always the case.

Consider the domain of basic cancer research, a field where we all hope that the publications are fully reproducible. A 2012 *Reuters* and *Science* story reveals the reproducibility crisis that is even present in that domain, with harsh consequences for new drug development [49, 48]. As head of global cancer research at a biotech company, C. Glenn Begley and his team of about

100 scientists reportedly tried to reproduce the findings of 53 published studies that they marked as 'landmark' publications [48]. The idea was to double-check the findings of these papers, before building further on the research in the search towards new drugs. The results were startling: only six studies held up and could be replicated, the other 47 (89%) could not [49]. As this does not necessarily imply fraud, but could rather be due to technical differences or difficulties, Begley and his team contacted the authors of the papers to discuss the contradictory findings. During a cancer conference, Begley reportedly sat down with the lead scientist of one of the papers that could not be replicated. The feedback he obtained is quite disillusioning, as described in his report of that meeting [49]: 'We went through the paper line by line, figure by figure. I explained that we re-did their experiment 50 times and never got their result. He said they'd done it six times and got this result once, but put it in the paper because it made the best story.' And this is not a unique story; scientists at Bayer reported a similar problem [360]: of the 47 oncology projects studied, only 20–25% could be reproduced, to come to the same previously reported findings.

This problem is not limited to one domain. A 2016 *Nature* survey among 1,576 researchers shows that more than 70% of them have tried and failed to reproduce the experiments of another group's experiments [31]. Even more astonishing is that more than half of them failed to reproduce their own experiments. The respondents were active in a variety of disciplines, such as chemistry, medicine, and engineering. Most however still trust the published results in their field, while at the same time 52% stated there is a significant reproducibility crisis. In some domains it is possible that small differences in the input (or context) can lead to different results. In the same *Nature* survey, actual fraud is only the ninth most reported factor that often contributes to the issue of irreproducible research [31]. The biggest reason that was reported by respondents to be the cause of this crisis was selective reporting and pressure to publish.

The competition among academics

For outsiders, the hypercompetitive nature of the academic environment is likely unknown. Some context to explain this. Academic researchers are (partly) evaluated on the number of papers they have published. The better the journal or conference that the researchers is able to publish in, the better the researcher is assumed to be: because of this largely accepted principle, more publications lead to easier access to additional funding, which then allows to hire more researchers and conduct more research. Hence the academic saying:

'Publish or perish'. This context has actually led to an exponential increase in the number of papers during the last century, but only a linear increase in ideas (as measured by unique phrases in article titles) [303, 146].

Academic publications are often 'nothing more' than a written paper, which details the work done by the authors. Merton, who studied the sociology of science, claimed that the basic currency for scientific reward is recognition [165, 297]. So we, academics, aim for publications in top journals, which often get cited by other papers, as to get recognition.

It is assumed that what is written is true, and that the research was conducted according to the rules of the research domain. When a paper is submitted to a journal, other academics will review the paper and comment, indicating what should be improved upon and whether the paper is acceptable for publication or not. Unclear results such as 'We conducted the analyses ten times and only once we got this really nice result' are more likely to be rejected than 'Our analysis led to this really nice result'.

Code of Conduct for Research Integrity

What are these rules that a researcher should obey by? The European Code of Conduct for Research Integrity [8] provides a framework on what principles to follow. This code is a reference document for all projects funded by the European Union, and as such is recognized by the European Commission as a reference for researchers and organizations. The four main fundamental principles of research integrity are [8]:

1. **reliability**: ensuring the quality of the research,
2. **honesty**: during the research, while reviewing, reporting, and communicating on the research,
3. **respect**: for colleagues and society,
4. **accountability**: for all aspects of the research, up to publication.

Clear misconduct therefore includes [8]:

- **fabrication**: making up results and publishing them as real,
- **falsification**: manipulating the process or data without justification, and
- **plagiarism**: using other people's work or ideas without proper accreditation.

At this point you might argue that these principles and practices are self-evident. Yet, when academics are not taught about these principles and related cautionary tales, they might be attracted to the appeal of tweaking the rules slightly, in order to get more publications. A 2013 *Science* study uncovered the existence of companies that put academic authorship up for sale [218]. Some authors would sell slots on their paper, another option is asking a company to write a completely new paper for you (typically review papers) on a certain topic. Once again, the motivation for academics to do so can simply be the recognition, but also to obtain promotion or even to avoid losing their job.

Unethical academic behaviour can also include more nuanced behaviour, for example, as an anonymous reviewer, asking the authors to cite your work [96], not disclosing weaknesses or own doubts about the research, or splitting up your manuscript into several submissions just to have more publications [185]. All *p*-hacking related practices, such as removing data points that don't match your hypothesis, or only reporting the measure that best suits your story, also fall in this category of unethical academic behaviour.

Not adhering to these truth-finding principles can have an actual negative impact on science, and society in general. The previously mentioned reproducibility crisis demonstrates the potential waste of time, effort, and funding. Academics therefore have a duty to take the research integrity principles to heart. Some real-life practices that could combat misconduct are the following [31, 218]. When an author is being added to a paper at the end of a reviewing process, an editor should ask why this is needed, and what the additional contribution of the author is. Reproducibility should be made easier, which goes hand in hand with asking for access to the data (or motivate why it cannot be shared, for example for confidentiality reasons), the code, and proper documentation for both. PhD supervisors should be encouraged and incentivized to properly train and follow up on starting researchers. Academic institutions should be aware that all of this requires time, and simply imposing more educational and administrative duties onto academic staff will possibly come at the expense of training, followup, and replication efforts.

The next cautionary tale illustrates how the quest for academic recognition can lead to a chase down a rabbit hole of unethical academic conduct, but first a discussion on how to evaluate a targeted advertising campaign.

Discussion 10

Targeted digital advertising is a massive industry which makes ample use of data and data science. Suppose you're working at an advertising agency that has been hired to run a display banner campaign to promote a new book on 'Data Science Ethics'. Your agency is known for its savvy and ethical use of data to target ads. You need to decide on how to report on the results of the digital campaign. There are three options: either report (a) how often a user clicked on the ad, (b) how often a user ended up on the landing page of the ad, or (c) how often the advertised discount code that is given in the ad was used in an online conversion (buying the book).

1. What are the scenarios under which these three metrics would be different?
2. Which metric seems most correct: clicks, visits to the ad's landing page, or online conversions?
3. Your analyses reveal that if the ad is shown in a gaming app for two year olds, the click through rates go through the roof. Yet, the number of visitors on the landing page remains low. What could be leading to this result? What would be reasons to show all ads within this app? What would be reasons not to?
4. Your analyses reveal that a webpage that leads to high conversion rates is a conspiracy theory page which includes hate speech. What would be reasons to show all ads on this webpage? What would be reasons not to?
5. What benchmark would you compare the results of your targeted advertising campaign with, so as to show that the additional data science effort led to measurable better results?
6. Your boss tells you to make a report that (a) includes only the metrics and benchmarks that make your campaign look as good as possible, and (b) not to report the webpages or apps where the ads have been shown. Would you do so?

5.4 Cautionary Tale of Diederik Stapel

Diederik Stapel was a well-respected professor in social psychology and dean at the University of Tilburg. His research attracted much attention, from both academics and the broad media, and he was awarded several prizes [56, 438]. When his data manipulation and complete fabrication of experiments were revealed [449], overnight he became what the *New York Times* reported as 'perhaps the biggest con man in academic sciences' [56].

In his research, Professor Stapel would often do survey experiments with human participants. Some of his widely reported findings were that meat eaters are more selfish and less social than vegetarians [120], and that a messy room makes people more aggressive [56].

The data on many of his experiments turned out to be fabricated. According to reports on the matter, he would develop a complete experiment, from theory and hypothesis to questionnaires, and would then make up participants' answers from experiments at schools that he would only have access to [56]. Reportedly, he would give the Excel files of survey results to PhD students to analyse, who would then jointly publish their findings and thereby contaminate their own PhD theses with fabricated data (without the knowledge of the PhD researchers) [449].

When fellow researchers began to notice something was fishy, the university was contacted and steps were taken. He has had more than 50 of his publications retracted [427]. Stapel later published an autobiographic book in which he explains what drove his descent from renowned professor to self-admitted perpetrator of data manipulation and fabrication [413]. He seemed to have been driven by power and the desire to be honoured, as well as the messiness of experimental results: the results from experimental data were too unclear, while the answers (for him) were clear, so he started manipulating the data accordingly. Stapel has expressed his sorrow, and apologized to his former PhD students and colleagues. A former PhD student of his reportedly stated: 'There are good people doing bad things, there are bad people doing good things.' She would put Stapel in the former category [56].

Total fabrication of experiments such as these is of course totally unacceptable. The practices started with 'simple' data manipulation, what one could call a gateway unethical data science practice. So remember this story when you are tempted into removing a dataset from your large-scale benchmarking study, or several data instances from your dataset, just to come to a more clear conclusion from your analysis.

5.5 Summary

Evaluating data science models in general is already a complicated endeavour. In this chapter we focused on the ethical aspects of the evaluation, which is structured in three parts: what to measure, how to interpret the results, and finally what to report. The ethical measurement requires that data science is conducted correctly, using all standards of the field. Simply not using an appropriate test set or a metric can lead to a deceitful story. On that point, remember Danny's limited test set and accuracy metric as choices to evaluate his stock prediction model. Ethical FAT data science measurements include the ones that were introduced in the previous chapters, such as fairness of the dataset and the model, privacy of the data and model subjects, and comprehensibility of the model.

In the second part, on how to interpret the results, we covered a range of practices that can lead to p-hacking. Even if there is no statistical test, such practices can still be unethical; consider for example choosing your data instances, doing input selection, and choosing evaluation metrics based on the test set. This is something that any decent data scientist would not do, as the interpretation of such results can lead to wrong conclusions and followup investments in the wrong direction, with potentially substantial negative impact on the business or society. Remember the investment fund inspired on Danny's work, which went under.

Finally, reporting ethically requires you to report transparently, the good and the bad. Explaining what you tried that didn't work out can be quite insightful for others as well, as it will guide future projects away from tried-and-failed avenues, or can inspire others to improve on the results. Ensuring reproducibility is a key aspect, which brings to light all the steps conducted, from input selection to model evaluation. Academic reporting is specific as it is guided by a quite unique competitive environment, and as it can have a profound impact on society. For an academic data scientist, it is therefore highly recommended to make the code of your work available, as well as the data (synthetic if needed), so others can reproduce your experiments and easily build further upon it.

6

Ethical Deployment

Chandler, Arizona, an American city with about 250,000 residents, wide and well-marked roads, that hardly ever sees any rain or snow [437, 283]. It's the city where self-driving car company Waymo, owned by Google-parent company Alphabet, is test driving its vans since 2017 [194, 379]. In the first two years, 20+ attacks were reported on the driverless vehicles [371, 379]. Tyres were slashed and rocks were thrown at the vans. Some driver even attempted to run the Waymo vans off the road, while another simply yelled at a van to get out of her neighbourhood. Why was this happening? Complaints filed with officials ranged from the risk of accidents to the potential loss of jobs that self-driving cars would bring about. 'They didn't ask us if we wanted to be part of their beta test' one of the residents that was issued a warning by the police for trying to run the vans off the road, stated to the *New York Times* [379]. Waymo reported that the attacks related to only a small fraction of all the tests that they have been conducting in Arizona, and place safety at the core of everything they do [379]. In this chapter, we'll discuss the ethical issues related to deploying a data science system. This starts with who gets access to the system. In the Waymo story: who gets to test the self-driving vans. Next, we'll discuss the ethics of treating people differently based on the provided predictions, honesty and oversight, and we'll cover some of the unintended consequences of data science deployments. The loss of jobs being an important one. Consider the social unrest in a country as the US when the more than 3 million drivers [25] and 3.6 million cashiers [439] will lose their jobs to automated data science systems, knowing the, little yet noteworthy, harassment of Waymo vans that already emerged in Chandler in 2018.

6.1 Access to the System

In the deployment stage, we need to discuss who has access to the system in which the data science is deployed. For a variety of reasons, system access can be limited to certain persons or regions. Often an ethical discussion is required

Data Science Ethics. David Martens, Oxford University Press.
© David Martens (2022). DOI: 10.1093/oso/9780192847263.003.0006

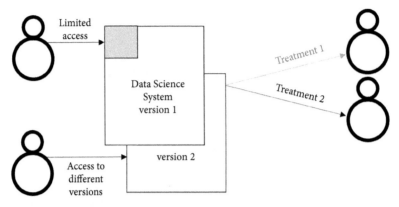

Fig. 6.1 Access to the data science-based system might be limited, be different for different persons, and lead to different treatments.

beforehand to determine who should be given access to what part of the system and why, as detailed next and illustrated by Figure 6.1.

6.1.1 Limited Access

A first reason is the sensitive and personal nature of the data, which requires organizations to limit the access to this data. Think for example of the payment data available at banks, medical records at hospitals, grades at universities, or tax information at the government. Next to obvious confidentiality, integrity, and access control measures, it is important to have a logging system that keeps track of each data access. If a banking employee looks up the payment history of a celebrity, he or she should be accountable, and be able to explain why this lookup was made. Whenever you think to have similar sensitive data, build in this logging functionality and make your employees aware of its existence and potential negative consequences when data is accessed without a valid reason.

Organizations may also decide to intentionally make *part of the system* inaccessible to certain users or in certain regions. A couple of examples. After the Google Gorilla problem was detected, the prediction of gorilla was simply turned off. Even two years later, Google Photos reportedly still remained blind to gorillas [402]. Is this right? Might investing more effort and resources resolve the problem? If it is an extremely hard problem to solve, should one communicate openly and transparently about the technical difficulties? Or is it simply a warning about the predictive accuracy of the sometimes overestimated deep learning models [402]?

Similarly, news feeds on news websites and social media platforms like Twitter and Facebook are often filtered to show the news that (is predicted to be what) you would want to read, thereby becoming separated from any news that might differ from your own point of view, in so-called echo chambers [36, 92]. Whether this is wanted is one thing, being transparent about the existence of such an algorithm, and how the algorithm works is another. One can choose to randomly add other news, not tailored to the user, as to break through the potential cultural and intellectual isolation that this might bring, or allow users to turn off the filter altogether.

Yet another example comes from the explainable AI domain. For a single prediction model and data instance, many different explanations can be generated, which all can seem equally valid, being the same length, all being counterfactuals, and so on. But one explanation might be more convenient for a business to provide to its customer than another. Let's consider the recruitment case. One explanations states: *if your gender were male instead of female then you'd be invited for an interview*, while another explanation states: *if your skill level of Python were advanced instead of intermediate, then then you'd be invited for an interview*. Both might be of similar quality in terms of length, providing a counterfactual, etc. Yet, the first one reveals a clear discriminatory bias in the model, the second doesn't. So a business owner might be more inclined to show the second explanation and conceal the first one. Solon Barocas and his co-authors pointed to this new power to decision makers [40]. Transparency in which explanations are shown when, both internally and to the model subject, can help to constrain this power.

Finally, the system could be *completely inaccessible* for some people. Any decision-making model that is built using smartphone data, is limited to data from people with a smartphone. Elderly people and people with lower income tend to be excluded from such a sample and hence might be negatively biased against. So this example shows that the lack of access to a smartphone can lead to a bias in the data gathering (cf. Section 6.6), but also in the deployment stage.

Some data science technologies have such major, possibly unethical uses, that the access needs to be controlled very stringently. Face recognition is such a technology. Based on a picture taken from any person, face recognition can match the face with a database of images of persons, if that person is in the database. This is a potentially great tool for law enforcement, to help solve crimes. But at the same time, there is tremendous potential for misuse: to curb individual rights by identifying persons (for example at peaceful protests), to stalk strangers, or to track unaware customers. Who should be given access to this software, and using what database of images, is a discussion that we'll get into in the discussion case of Section 6.3.

The large degree of freedom to decide who gets access to the system creates a new form of power to companies and governments using data science. Once more it is important to be transparent in the choices made, with motivation for the choices and ethical considerations that were made.

6.1.2 Different Versions for Different Persons

So far we have discussed how access to a data science system might differ between users. Also different versions of a certain data-driven app or software could be provided to different users (or regions). A user may actively choose to install or use a different version, for example one with ads and targeted content, or a paid version with more control on what data is used and without ads. Important ethical considerations come into play when it is the company itself that makes the decision for the user which version he or she will get. Often these are related to complying with policies and legislation.

A widely reported related practice was that of ride sharing services company Uber, where in its early days two versions of the app were created [221, 269]. According to the reporting, Uber was dealing with fraud, where some Uber drivers would create fake accounts and request rides, which they would then accept (after which they would cancel that same ride again). Uber was incentivizing taking more rides, so the fraudulent behavior would earn the drivers more money [221]. By erasing the phone, and creating a new fake account, this could be done over and over again, without being able to identify the phone. Uber started to 'fingerprint' iPhones, so that the phone would get a unique identifier, even after it was erased. By doing so, Uber could detect users who repeatedly use, erase, and re-use a phone to commit fraud in their system. But this practice was troublesome in itself, as fingerprinting was against Apple's privacy rules [221]. To circumvent this policy issue, Uber is said to have created two versions of its app: one for the broad public, and one specifically for Apple employees at Apple headquarters. By geofencing Apple's Cupertino headquarters, Uber would know which iPhones probably belong to Apple employees. A different version of the app could then be offered to those Apple employees, so as to disguise that Uber was identifying and tagging iPhones even after the app had been deleted. The practice was later stopped, after Apple's CEO Tim Cook reportedly asked the CEO of Uber to do so, or face the consequences of the Uber app being kicked out of the Apple's App Store [221]. In this case a different version is provided to cover up a rather unethical practice, far from the transparency criterion we aim for, even though it seemed to address a serious fraud issue.

Companies might also decide not to provide access to certain users, because the system might violate the regulations of the given users. The day after the European GDPR came into effect, many non-European websites simply blocked the content of the website to European visitors [43]. This of course can be a valid business decision, so as to avoid legal and technical costs to comply with the regulation (or buy time to do so). Two months later, about a third of the 100 largest US newspapers reportedly blocked their content for European visitors [410, 333]. This raises ethical questions as it concerns news: some European users rely on these news sites for their daily news; others would like to read a specific article that is only accessible on that news site. Rather than taking the effort to comply with GDPR, and in the meantime also strengthening the privacy of the non-European visitors, many news sites just chose to take the arguably easy road of blocking the content to European users.

6.2 Different Treatments for Different Predictions

6.2.1 Data-driven Price Differentiation

Whoever regularly books airline tickets will know that ticket prices change over time. Even more: airline passengers on the same flight are likely to have paid different ticket prices, based on available seats and time of booking for example, but also on whether a business or economy class was chosen, and whether the passenger is registered in any loyalty programme. This practice has become quite common and is an example of differential pricing. A related example is different (costlier) hotel rooms being shown to Mac versus PC visitors [292]. Varian found that offering different versions of a good at different prices increases the overall welfare, as measured by consumer surplus and producer surplus, if it allows to serve new markets that would not be served without versioning [448]. Providing a student discount for example, might bring a new set of consumers into the market.

What we focus on here is different treatment of persons based on data-driven predictions made on those persons. In the airline ticket example, a prediction of willingness to pay could lead to different prices to charge, or different ads to show (economy versus business). Such a prediction can be quite crude, for example only based on the country or zip code of a person [281]. In a credit scoring context, risk-based pricing will link the interest rate to the predicted credit risk [446]. Such practices make economical sense, as a larger risk leads to a larger expected loss, due to a higher probability that the borrower will not repay the loan, and hence should be compensated with a larger price.

So what is the ethical issue here? Once more it is about transparency and fairness. The practice could lead to price discrimination based on sensitive attributes. We've already covered the dangers of using location (such as zip code) in predictive models, which could negatively impact certain racial or income groups, cf. redlining. Other socio-demographics such as age and gender are reportedly to likely correlate with ability and willingness to pay [448]. Unsophisticated, vulnerable, and technological unsavvy consumers are arguably less likely to evaluate the degree of price discrimination that applies to them [253], and will likely not know how to adapt to counter potentially negative price decisions on them. This same reasoning applies more broadly to any prediction-based treatment, beyond pricing, such as information provided or services offered. The second fairness issue is related to privacy: is there any personal data used to make those predictions? If so, this begs all the previously addressed questions: How was the data gathered? Did the user provide consent for this? Was this the initial purpose of the data gathering? How was it preprocessed? Can a user be re-identified based on the data? And so on. The third issue is transparency. Different treatments require transparency for the consumer: if the condition of overall welfare increase is met, there will be a consumer surplus. Making this surplus, which is triggered by different treatments, transparent would therefore we warranted. If the different treatment remains concealed, there is likely an ethical argument against it to be made which led to the decision to not disclose the practice. Having the difference subsequently revealed by an outside party, such as a journalist, can lead to even more ethical headaches and reputational damages. So have the ethical discussion: What is the effect on consumers, who benefits, who doesn't? Is it right? For example, if you find that your Apple customers tend to be wealthier, is it right to charge them more for your services, potentially allowing you to charge the others less? There might be reasons to justify this, as it seems similar to the progressive tax system that many countries have. But as a Mac user, would you consider such a practice fair? What if a Mac user is offered a more expensive version or treatment first, but now with an explanation why this treatment was provided (Mac users tend to choose the more expensive treatment), and an option for consumers to not use data at all which would lead to no differential treatment? This ethical discussion will lead to your data science equilibrium and a decision on whether your customer treatments should be different across different predictions.

6.2.2 Behavior Modification to Make Predictions True

Galit Shmueli takes this issue a step further, by focusing on how Internet platforms not only have behavioural data to make a range of predictions, but also
have the tools available to change our behaviour [397]. She makes a compelling
case that this combination incentivizes the platforms to 'make the predictions
true', by using behaviour modification techniques. Why would they do so?
Well, the predictions can be sold, in what Shoshana Zuboff describes as 'prediction products' [481] to businesses as advertisers, insurance companies, and
political consulting firms in so-called 'behavioural futures markets'. Obviously,
the more accurate its predictions, the better for their business of prediction
products. Shmueli provides several hypothetical cases on how such behaviour
modification might work. The first is that of an insurance company that seeks
to attract new low-risk customers. An Internet platform might be offering such
risk predictions, based on the ample behavioural data that they have access to.
The behaviour of the low-risk predicted users can be pushed towards that prediction by encouraging them not to use the app while driving and not showing
ads for alcoholic beverages during working hours. The opposite can be done
for the high-risk predicted users, potentially even explicitly pinging them with
messages while they are driving. Another intriguing scenario that Shmueli
provides is that of an election campaign. Once more the hypothetical platform
has all the tools to push the voting behaviour towards the initial prediction,
for example by simply showing more positive messages, ads, or search results
for the candidate that the person is predicted to vote for, and negative ones for
the opposite candidate.

As apparent as the incentives for the platforms to do so may be, the ethical issues are abundantly arguing against such practices, as also described by
Shmueli [397]. First, it would unfairly treat some consumers, potentially even
towards causing harm to them (cf. the insurance case) or society (cf. influencing an election). The consumers are even blind to the potentially damaging
practice, both in terms of use of their data for this goal and in the behaviour
modification actions, unless the platform would explicitly reveal them the existence of these mechanisms. The business customers of the prediction products
would similarly be in the dark, and will be overestimating the predictive accuracy of the prediction products. Even the platform itself might be unaware of
the behaviour modification, for example when the predictions are automatically included in further personalization/modification tools: predicted to be
high risk? Well, then the algorithm finds that you are more likely to be interested in this whisky-drinking-at-work ad. This reasoning argues once more for

explicitly thinking through and discussing the ethical consequences in terms of fairness and transparency of your data science choices.

6.3 Cautionary Tales: Censoring Search and Face Recognition

6.3.1 Google Search in China

The deployment of Google Search in China can be described as a complex dance between two giants with seemingly clashing ethical values. Google's mission is to 'organize the world's information and make it universally accessible and useful' [172], and immediately sheds light on the tension between Google Search and China. Before discussing this tale, we need to understand why Google wants to be active in China, and why China would welcome Google. The huge market size of China is of course a first motivation for wanting to be active in that country. As shown in Table 6.1, whereas China's online population in 2006 was well below that of the United States, ten years later it became nearly three times the number seen in the United States. The ability to serve this huge pool of Chinese internauts and potential customers is of course an immense business opportunity.

A second, arguably even more interesting motivation for Google is provided in a reportedly leaked speech by Google's head of search, given in 2018 [159]: 'China I think is one of the most interesting markets, arguably the most interesting market in the world today. Just by virtue of being there and paying attention to the Chinese market, we will learn things, because in many ways China was leading the world in some kinds of innovation.' The success of Chinese technology companies outside of China, such as TikTok and Alibaba, strengthens this argument.

The motivations for China to welcome Google are put forward in a *MIT Technology Review* article [394]: to obtain improved results for search queries (certainly when looking for international information), the ability for Chinese companies to partner with Google and thereby have access to the technological

Table 6.1 Online population in the United States and China [428, 429].

Country	2006	2016
United States	206 million	276 million
China	138 million	734 million

expertise and prestige of the company, and the legitimation of the Chinese Communist Party.

So in 2006, Google launched google.cn, a censored version that would comply with Chinese laws and policies [94]. Google acknowledged the ethical balancing act it had to perform: 'While removing search results is inconsistent with Google's mission, providing no information (or a heavily degraded user experience that amounts to no information) is more inconsistent with our mission.' [94] In the search results, Google would additionally reveal to the user when certain results have been removed. This notification is an important victory for transparency, and reportedly wasn't popular with the Chinese regulators [394]. Baidu and other Chinese search engines soon followed the same practice.

In 2010, Google decided to remove the censorship from the results after a major hacking attack of the company that originated from China [44, 394]. This hack seemingly was intended to get access to the gmail accounts of Chinese human rights activists [117]. After failing to reach an agreement with the Chinese government over its censoring, Google search was banned in mainland China. In that year, China also blocked access to platforms as Facebook and Twitter, after rioting in the region of Xinjiang, in a move that would lead to what is now known as 'The Great Firewall of China' [394]. In the absence of Google, in 2016, Chinese search engines started to remove the notifications that results were being censored [394].

Then, in 2018, The Intercept revealed that Google was working on a censored search engine for China again, named Dragonfly [6, 157], which was designed to block information related to democracy and human rights, and filter out websites as BBC and Wikipedia [158, 322]. After quite some uproar and internal complaints from the company's privacy team, Google reportedly 'effectively ended' the project [158]. In July 2019, Google officially confirmed that Dragonfly was terminated and that it had no plans to launch search in China [6, 322].

What are the lessons to be learnt from this cautionary tale? First, that such ethical discussions have no easy answer. Not only is this a balancing act between ethical and business considerations, one can also make an argument why ethically it is worthwhile to enter China, as it improves transparency, could change the system from within, and is in line with the mission of providing information to everyone. The roller coaster that was the Google Search censoring practice in China is evidence of how such a delicate decision can quickly change based on the time and events that take place. Second, it demonstrates how transparency was a force for good in this tale: in revealing and notifying users that some results are being censored, but also in revealing that the censorship was stopped and why. And third, the remarkable impact that employees and society can have in the decision process, as shown by the employees of Google making a stand and demanding to drop the DragonFly project [157].

6.3.2 Access to Face Recognition Software

Face recognition software is and likely will always be an ethical focal point, surely when discussing who is allowed to use such software. An interesting case comes from Clearview.ai, as already introduced in Section 3.4. This tech company offers face recognition software and provides a link to online available profiles of the matched person, such as on Facebook and Linkedin, after uploading an image of that person [209]. One of the ethical issues related to this service, and more generally any face recognition software, is who should have access to this technology? Only law enforcement agencies, or also security departments from private industry? Should this be limited to countries that respect human rights, or should it be available everywhere? Should certain citizens also be allowed to use the app, for example registered private detectives? Are the (potential) investors of Clearview.ai allowed to use and verify the app? Are the Clearview.ai employees allowed to use the software for any purpose? These are the kind of ethical questions related to limited access that a company should think through before launching such an app. Given the power that comes with deciding who is granted access, it is important to have a transparent process in place to decide on this, and report transparently on the decision and decision-making process. Finally, the policies should be transformed into demonstrable and effective measures, with potential negative consequences for whomever violates these policies, in short: be accountable.

6.4 Honesty and DeepFake

'Not that you lied to me, but that I no longer believe you, has shaken me.' This quote by Friederich Nietzsche [326] summarizes the potential negative consequences of the data science technology DeepFake, where we might not believe footage in videos or recorded speeches any more. DeepFake is a technology based on deep learning to create real-looking yet fake videos. One of the first widely known DeepFakes movies showed former US president Barack Obama pointing out the dangers of false news. In this video, titled 'You Won't Believe What Obama Says In This Video! ;)', Obama opens with: 'We're entering an era in which our enemies can make it look like anyone is saying anything at any point in time. Even if they would never say those things.' Later in the video it becomes clear that comedian Jordan Peel is doing the actual talking, and that the video is a fake video produced by BuzzFeed [75]. Disinformation campaigns are not new, and predate the digital era. The rise of digital and social media have made it easier however to manipulate and share such intentionally deceptive information [16, 291]. What's new in DeepFakes is the ease to change video materials, which are convincing and are almost indistinguishable from real, authentic material [179].

The most popular type of DeepFakes are videos in which the faces are swapped, while retaining the expression, pose, and background area of the original image [366]. One popular open source tool is FaceSwap [366],[1] created in 2017. Technically, the tool uses two deep auto-encoders, where the encoder part is shared. This encoder will map an image to a shared latent representation, and each separate decoder will then disentangle the facial identity from the facial expression. The result is a swapped face: the original person's identity, but the imposter's expression and pose. This inexpensive and relatively easy to use technology can be applied to publicly available video (or voice) recordings to impersonate a person.

Let's first list some potential legitimate uses of this technology. In entertainment, the technology has been used to include characters in movies where the actual actors cannot play. For example, a younger version of the actor, or when the actor passed away, or simply is unavailable. Well-known examples include an ESPN commercial [214] and several *Star Wars* scenes [418]. The online community has recreated numerous movie scenes where actor Nicolas Cage is inserted into the role of other actors [372]. Belgian special effects expert Chris Ume has also created several short DeepFake movies of Tom Cruise. He stresses that creating such lifelike movies is not that straightforward to do, but he also made clear that the use of DeepFake technology in movies is here to stay [452].

Potential wrong uses are unfortunately ample as well. Also in the entertainment business, movies are being created with deep learning where the actor does not want to play in. Celebrities such as Jennifer Lawrence and Emma Watson have been inserted into fake porn movies [179]. Revenge porn with DeepFake can also do much harm [179]. Fake videos can be made, and used to blackmail persons, for example because of embarrassing or criminal fake footage that has been made. Deepfakes are not limited to video, and can also be used to impersonate a person's voice. In 2019, the CEO of a United

---

[1] https://faceswap.dev/

Kingdom-based energy firm became victim to a suspected DeepFake speech scam, and lost 225,000 Euro in the process [419, 179]. He/she thought to be speaking with his/her boss, the CEO of the parent company, who asked to urgently transfer money to a Hungarian supplier [419, 179]. In political campaigns, fake news stories with fake statements can be used to put the opponent in a bad spotlight. Or similarly, political actors can simply deny being in a negative recording, simply claiming it is a DeepFake.

A 2021 experiment by Hwang et al. looked at the effect of a DeepFake video on disinformation, with 316 Korean adult participants [475]. Participants were shown a fake news article on Mark Zuckerberg, the CEO of Facebook. One group were additionally shown a DeepFake video. The results showed that including a DeepFake video in a disinformation message led to significantly greater persuasiveness, credibility, and intent to share the message, compared to the textual message only. So DeepFakes seem to indeed be a useful instrument to foster disinformation. The study also showed that the effect of the disinformation could be countered by literacy education, where participants would be educated about the definition of disinformation, shown some examples, and informed about the consequences of disinformation [475].

DeepFake technology ignites the search for other methods to detect fake videos. Jiameng Pu and her co-authors looked at existing DeepFake detection algorithms (available at that time, in 2021) on a dataset of 1,869 videos from Youtube and Bilibili [366]. They found that none of the methods obtained a high detection rate on these videos, defined as having a F-score over 90%. Watermarks or blockchains might further help to verify if a video is authentic or not [179]. As the experiment of Hwang already demonstrate, literacy education is an important non-technical answer to malicious DeepFake videos and speeches. Or as Prof. Lyu from the University of Albany puts it: 'Awareness is a form of inoculation.' [179] Regulations and code of conducts also start to emerge to control DeepFakes. For example, a US federal law aimed at Deep-Fakes [89] was passed in 2019, where the government encourages research of DeepFake-detection technologies; while Texas and California both passed laws to prohibit DeepFake videos in the context of elections [179].

6.5 Governance

The increased importance of data science and data science ethics for companies comes with a need for proper governance and oversight of these aspects. This includes the following three steps: set up an ethical oversight committee, establish a data science ethics policy, and put practices and processes in place to implement the policy.

Set up an Ethical Oversight Committee

An Ethical Oversight Committee can be put in charge of formulating the ethical principles that the company wants to adhere to, in all stages. The policy with the principles should guide ethical questions that might arise, such as: what personal data to keep and for how long (gathering), should we predict pregnancy (preprocessing); what explanation to provide our rejected customers, if any (modelling); how to assess fairness of our credit scoring models (evaluation); or who gets access to our face recognition software (deployment)? Once the policies have been set up, the committee can also be charged with the reviewing and approving/rejecting of potential data science uses [251], and guide additional tooling and training practices.

The committee members should be knowledgeable, be able to spend the required time, and be recognized within the company. First of all, these should include representatives from all stakeholders, be it from business, data, and model subjects (focus on those negatively impacted) and data scientists. The more diverse, not only in role and professional background, but also in terms of gender, culture, and age [203], the more likely that potential issues will be uncovered [251]. Second, senior management should be involved to ensure that the committee is given proper resources and time, but also to demonstrate the commitment to, and importance of, data science ethics to the company. External members and thought leaders can additionally help to provide additional insights [251]. The committee should foster an open discussion culture, where all angles are considered and all opinions can be voiced, leading to a chosen golden mean. Finally, any employee should feel comfortable with raising potential issues to the committee, anonymously if needed. The committee should report transparently on who the members are, how they were chosen, what decisions have been made, and what their role is.

An interesting example is Facebook's Oversight Board, sometimes refered to as Facebook's 'Supreme Court' [45, 312, 219]. The board was set up in 2020, after Mark Zuckerberg reportedly met with Harvard Law School professor Noah Feldman in 2018, who argued for the creation of a board [191, 425]. Some of the intended goals are to include fairness, create transparency into the systems, have independent oversight, and increase accountability [482]. More formally, the purpose of the board is reportedly 'to protect free expression by making principled, independent decisions about important pieces of content and by issuing policy advisory opinions on Facebook's content policies' [339].

Users can hence appeal decisions on content moderation to the Board, which may then take up the case and make an independent judgment. Ensuring independence of course is an important component for this board. One way is the inclusion of outside experts as board members. The 2021 members include a Nobel Peace Prize Laureate, journalists, academics from various disciplines and the former prime minister from Denmark [340]. To ensure further independence, a trust was set up to safeguard the independence of the board and to ensure its effective operations. Additionally, the decisions of the board are said to be final [339, 425, 219]. These responsibilities and relationship with Facebook are described in a charter, which enhances the transparency of the board and its decisions [341]. A well-known case that the board looked into, was the suspension of the account of former US president Donald Trump following the 6 January 2021 riot at the US Capitol. The board upheld the decision, but argued that Facebook should reassess its decision of indefinitely banning the former president from the platform, and Facebook would have six months to do so [45, 219].

Establish a policy

An important governance issue is describing what is considered ethical. A set of key principles that relate to data science practices is the culmination of the relevant data science ethics aspects that we discussed in this book. Such a policy can provide a guide to employees, be used in training, and provide structure to remedy potential violations that might occur [116]. These principles are distilled from discussions on the topic. This conversation in itself is of tremendous value, as it coincides with thinking through the potential uses and misuses, and impact on all potential stakeholders, which may be seeded by the various Discussions, opening stories and Cautionary Tales provided in this book. Of course, this requires input from all stakeholders, and can be driven by the Ethical Oversight Committee. The principles to adhere to are likely different according to the sector and size of the company at hand, and should be aligned with the values of the company. So have the discussion and be transparent in the chosen golden mean, between excess and deficiency, which will then define the data science practices. The AI principles of Google, made public in a 2018 blog post authored by CEO Sundar Pichai, provide a nice example. The first (of seven) principles is headlined [350]: 'Be socially beneficial', yet another is 'Incorporate privacy design principles'. Applications that Google will not consider, according to the blog post, include technologies that cause (or are likely to cause) harm, and weapons that cause injury to people [350]. Yet another relevant set of principles are provided in the ACM Code of Ethics and Professional Conduct [116], which includes such principles as avoiding harm, trustworthiness, fairness, and privacy.

Implement the policy

The implementation of the policies firstly requires that employees are made aware of the existence of the policies. This can be achieved by including them in initial policies that new employees sign off on (similar to IT policies), and training tailored to the role and background of the employee to create an ethical mindset [203]. To ensure accountability, effective and demonstrable measures should be taken to move from theory to practice. The numerous data science concepts and techniques that we covered so far can provide an inspiration to that matter. Employees should be made to understand the reasons for the practices, so they don't consider the policy as just another document to sign off on. Communication is key to that effect, and can take many forms, from presentations at company events to blog posts. Similarly think about how you will communicate towards your external stakeholders about your principles and practices, and how they can reach out to you for wanted feedback (e.g. explanations for predictions) or to provide comments. Inspiring examples from Facebook on that matter are their blogs on 'Privacy Matters'[2] and 'Hard Questions,'[3] and the 'Why am I seeing this post'[4] feature.

6.6 Unintended Consequences

Imagine that a computer needs to figure out how to eradicate cancer from the world. After the analysis of massive volumes of data, a data science formula is spit out which states to kill everyone on earth … This is a scenario that has been put forth by Bossmann in 2016 at the World Economic Forum to illustrate the issue of unintended consequences of using AI [60]. Although the example surely makes the point, many other more realistic situations exist where a data science model comes with unintended consequences, which typically fall in one of the following categories: either the data science model behaves differently than intended, illustrated by the 'kill all mankind' formula, or the impact on humans is unintended, where some humans suffer (economically or even medically) as collateral damage.

---

[2] https://about.fb.com/news/tag/privacy-matters/
[3] https://about.fb.com/news/category/hard-questions/
[4] https://about.fb.com/news/2019/03/why-am-i-seeing-this/

6.6.1 Data Science Model Behaving Differently than Intended

The first group, where the data science model behaves differently than expected, often involve relatively simple bots interacting with their environment, resulting in complex, unexpected overall behaviour.

On the online encyclopedia platform Wikipedia, changes to articles are being made by both humans and automated bots. The bots perform relatively simple tasks, such as spell checking, enforcing bans, creating inter-language links, and identifying copyright infringements [324]. Tsvetkova et al. investigated the behaviour of such editing Wikipedia bots, during the period 2001 to 2010 [436]. They surprisingly found that some of these bots were undoing each others' work, by unediting the changes made by other bots. The most common fight was that of creating and modifying links between different language versions of a Wikipedia page. What is so interesting is not just that these bots had these kinds of fights, but that it often even went on for years [436].

Another example comes from the financial world, where algorithms have been used for a while for high-frequency trading. It has been reported that algorithmic trading accounts for about 70% of all US equity volume [417]. Very short-term algorithmic trading is known as high-frequency trading, where (data science) models and algorithms try to spot opportunities in the stock market, often in a matter of microseconds. These trading models reportedly have little interest in a company's fundamentals or fate [417]. The behaviour of such high-frequency trading algorithms has been studied in relationship to flash crashes, where the stock market suddenly falls sharply and regains its value, in a very short time period. And even though the causing role of such algorithms in flash crashes has been questioned, they can magnify the impact of such unexpected events [433, 352].

A final case is that of 'Tay', the so-called racist Twitter bot [216, 391]. Tay was developed by Microsoft to entertain 18- to 24-year olds in the US on Twitter while experimenting with conversational understanding [263, 337], and was released in March 2016. Only 16 hours after its start, it was already shut down because of the offensive, racist and sexist tweets it was writing [391, 337]. The Twitter profile of Tay stated: 'The more you talk the smarter Tay gets' as it would learn from the interactions. Microsoft explained that Tay was attacked by so-called trolls, who asked racist and sexist questions in an attempt to bait Tay to say nasty things [391, 359]. And even though Tay was reportedly intended to create positive experiences, and extensively tested, the incident revealed a vulnerability in the bot. The behaviour clearly was unintended, and Microsoft arguably did the right thing: it apologized, turned off the bot, and removed the offensive tweets [359, 337]. This cautionary tale teaches us the lesson to think through the potential misuse by adversaries, already in the design phase, and to consider the potential offensive language that might emerge. Microsoft would later release 'Zo', a new bot version that would shut down when conversations would steer towards sensitive topics as religion and politics [391].

6.6.2 Impact on Humans Is Unintended: Job Loss(?)

Many tasks that are currently done by humans will likely be replaced by intelligent robots and data science systems in the future. Next to the positive implications, such as increased productivity, safety, and even overall wealth, there are also highly disruptive and negative consequences. A loss of jobs being a major one. Think for example of the trucking and driving business. With the advance of self-driving cars, trucks, trams, and trains, these drivers might very well find themselves out of a driving job in the near future. The impact of automation on jobs was even a focal point for 2020 US presidential candidate Andrew Yang. In a 2019 opinion piece in the *New York Times*, titled 'Yes, Robots are Stealing Your Job', he described the daunting task of trying to offset the lost jobs by spurring entrepreneurship and attempting to create jobs: 'We were pouring water into a bathtub with a giant hole ripped in the bottom.' [472] Much has been written about this topic, and predicting what the future will hold is by definition speculative. But let's have a look at what is different now from similar historical scenarios, what jobs are at risk, and what potential solutions would look like.

The fear of job loss due to the introduction of innovations in the economy is not new. In the US for example, about 5.6 million manufacturing jobs were lost between 2000 and 2010, 88% of which are attributed to automation and an increase in productivity [201]. In an almost 100-year-old, yet still very relevant and intriguing essay, John Maynard Keynes writes that we are being inflicted with a new disease, termed technological unemployment, which he defines as 'unemployment due to our discovery of means of economising the use of labour outrunning the pace at which we can find new uses for labour' [240]. He states that the economic growth brought by innovations is accompanied by hardship: 'We are suffering, not from the rheumatics of old age, but from the growing-pains of over-rapid changes, from the painfulness of readjustment between one economic period and another.' Even earlier, in the 16th century, there is the tale of William Lee, who invented a stocking frame knitting machine [3]. In an endeavour to seek patent protection, he travelled to London to meet with Queen Elizabeth I. But she refused to grant him the patent, in fear of how it might effect hand-knitting employment: 'Consider thou what the invention could do to my poor subjects. It would assuredly bring to them ruin by depriving them of employment, thus making them beggars.' [3, 151] This is often a first reaction: holding back the innovation to save employment. But history shows that such innovations are unstoppable, and knitting is done largely by machines these days.

History also reveals that innovations actually lead to a net job *creation*. An MIT study on 'The Work of the Future' [26] lays out three reasons for this phenomenon. First, workers become more productive in areas that are not automated. Think of an architect who spends less time on drawing his or her designs, and more time on the creative process. Second, an increase in productivity increases the total economic pie. More income leads to more spending, which leads to more work. And third, new jobs emerge because of the innovations. It's estimated that more than 60% of the 2018 jobs were not yet invented in 1940 [26]. The modern data scientist role being an obvious one. But also many non-high tech jobs are new, such as mental health and fitness coaches. So can we conclude that we shouldn't be worried about the losses of jobs? Well, we probably should, as the net gain job increase is not evenly distributed across all demographic groups, this process of losing your job can be quite disruptive to one's personal life, and things might just be different this time. Let's first zoom in on who is about to lose their job.

The evident example are drivers, who will be impacted by the introduction of self-driving cars. Imagine the social unrest if millions of drivers find themselves out of a job from one day to another, knowing the anecdotal evidence from the town of Chandler. These drivers tend to be less educated, and immigrants with a language barrier [26]. Another MIT study estimates that the deployment of self-driving cars will take time and will happen gradually, while the additional work of loading, unloading, and maintenance will remain [265]. That would surely soften the disruption for the millions of impacted jobs. But the job loss will go beyond this often-discussed job of drivers. Several studies make estimates of potential job losses over the next decade(s), which all point to substantial numbers, up to around 30–40% for advanced economies.[5]

---

[5] Frey and Osborne from Oxford University for example find that 47% of US jobs are at a high risk of becoming automated, perhaps in the next decade or two [151]. A 2019 OECD study estimates that *only* 14% of jobs are at risk of automation, while 32% of the jobs might be 'radically transformed'. Yet another study, by McKinsey & Co, finds similar numbers, with 3 to 14% of the global workforce that will need to switch jobs by 2030, but they find more profound changes for advanced economies due to higher wages and hence incentives for automation [280]. For the US and Germany for example, the numbers rise to around 30%.

The types of jobs at risk include not only jobs that focus on routine tasks, but also non-routine manual tasks such as driving and surgery, and even non-routine cognitive tasks such as legal and financial services or fraud detection [151]. A 2020 study by the World Economic Forum lists the roles that are foreseen to have the most growth and decline by 2025, the top three of which are listed in Table 6.2. Data science related roles are on the uptick, and the role of data entry clerk is most at risk. A related 2020 study by the US Bureau of Employment estimates that about 6 million jobs would be added (net) in the US by 2029 [442]. They also rank occupations according to most forecasted growth and decline from 2019 to 2029 (see Table 6.3), in which cashiers top the list as the occupation with the steepest foreseen decline.

These tables already indicate that not just low-income low-skilled jobs are to be lost. Even more, there will be a shift from middle-income employment to low-income service employment [151]. This projected shift continues an already happening trend: between 1980 and 2005, the number of hours spent in low-skilled service jobs rose by more than 50% among non-college degree workers [27].

So jobs are going to be lost, the content of many jobs will change, and new jobs will emerge. Why should we worry about this? First, if the shift from middle-income to low-income jobs is not addressed with proactive efforts,

Table 6.2 *Roles* most and least frequently identified in the World Economic Forum Future of Jobs Survey 2020, as to be in demand by 2025 [390].

Most growth	Most decline
Data analysts & scientists	Data entry clerks
AI and machine learning specialists	Administrative and executive secretaries
Big Data specialists	Accounting, bookkeeping, and payroll clerks

Table 6.3 US 2019–2029 forecasted *occupations* with most job growth and decline, according to US Bureau of Employment [441, 440].

Most growth	Most decline
Home health and personal care aides	Cashiers
Fast food and counter workers	Secretaries and administrative assistants, except legal, medical, and executive
Cooks, restaurant	Miscellaneous assemblers and fabricators

existing inequalities will likely deepen [390]. Second, what is particular to the current situation is the impact on non-routine jobs, such as the auditor or radiologist, as well. Although these jobs might not disappear altogether, their content surely will, with more focus on being a bridge between the machine and the model subject (patient, or company being audited for example). Given that many people's employment will be negatively impacted, three solution strategies are consistently put forward in the literature: training, some variant of a universal basic income, and learning to live with the freedom of not having a job.

Reskilling unemployed workers is an obvious direction to take. The skills of the future are likely those that are hard to be automated and focus on collaborating with machines. These include applying expertise, interacting with stakeholders,[6] applying emotional and social skills, and advanced cognitive skills such as creativity [280]. Such continuous learning goes beyond the training of students in schools and universities, and businesses should allow their employees to upgrade their skills on the job. Technology companies that leverage new technology and thereby disrupt the global workforce could even lead the effort to retrain affected workers. In a 2017 Harvard commencement speech, Facebook's Mark Zuckerberg discussed this topic and proposed the following: 'And as technology keeps changing, we need to focus more on continuous education throughout our lives. And yes, giving everyone the freedom to pursue purpose isn't going to be free. People like me should pay for it, and a lot of you are going to do really well, and you should, too.' [196] Governments also have an important role to play, as they can provide incentives (be it in terms of subsidies, tax deductions or setting up training programmes) for reskilling and upskilling workers [390].

A second solution approach is to provide a universal basic income. This can come in various forms, from a guaranteed income per adult to unemployment insurance benefits. As jobs will be lost, a guaranteed basic income creates the time to reskill and provides a safety cushion to deal with the disruption caused by technology. In his 2020 US presidential campaign, democrat Andrew Yang proposed a monthly guaranteed income of 1,000 US$ per adult. Well-known tech entrepreneurs such as Mark Zuckerberg [196], Elon Musk [93], Jack Dorsey [81], and Larry Page [54] all advocated a similar idea. A political and moral question is who should pay for the additional spending: everyone

---

[6] The need for explanations of predictions, as seen in Section 4.4, once more becomes obvious in this context of being a bridge between the data science model and the affected stakeholders.

in society, mostly the wealthy, the companies whose innovations made the workers unemployed, or the unemployed persons themselves?

A final, more upbeat, strategy is to embrace the idea that machines will take over our jobs and learn to enjoy the jobless freedom that comes with it. Chinese tech entrepeneur Jack Ma stated the following: 'I think people should work three days a week, four hours a day … I think that because of artificial intelligence, people will have more time to enjoy being human beings. I don't think we'll need a lot of jobs. The jobs we need are [ones to] make people happier. People experience life, enjoy [being] human beings.' [143] As pleasant as this idea might be, the prospect of 'simply' being a stay-at-home dad or mom, an artist, or author is actually quite hard for most of us, and is arguably not held in high esteem by society as much as being a successful CEO or data scientist. Former US president Barack Obama phrases it as follows: 'we have to make some tougher decisions. We underpay teachers, despite the fact that it's a really hard job and a really hard thing for a computer to do well. So for us to reexamine what we value, what we are collectively willing to pay for – whether it's teachers, nurses, caregivers, moms or dads who stay at home, artists, all the things that are incredibly valuable to us right now but don't rank high on the pay totem pole – that's a conversation we need to begin to have.' This idea echoes what Keynes had put forward in his 1930 essay: '[W]e have been trained too long to strive and not to enjoy' [240].

In conclusion, data science may have tremendous positive effects on our economy and society, but there will also be tremendous disruption to the global workforce due to the automation that comes with it. You, the reader, should be aware of the consequences of your work on many people's employment, and should not shy away from being part of the ongoing discussions on this matter.

6.7 Summary

Once the data science model has been properly built and evaluated, still ethical issues might emerge in the deployment stage. This chapter looked firstly at who would get access to the system. Limited access is surely warranted when dealing with personal and sensitive data—think of a hospital or a bank that needs to decide on who gets access to what data under what circumstances. But the choice to limit the access can yield power to the one granting the access: which information to recommend or which explanation to provide for example. Another ethical decision to make is who gets to use the deployed system. Some persons might simply not have the technical or financial means to use your system—think of elder persons without smartphones. This might be a conscious choice of course. You might also intentionally disallow access to certain persons or companies, as motivated by face recognition technology. The distinction in who can and is allowed to use your system can also argue for having different versions for different persons.

Next, we looked at the different treatment of persons based on the predictions made by the data science system. Differential pricing is a quite common and accepted practice in domains as the travel industry. When charging different prices based on the prediction of a data science model, once more the transparency and fairness issues emerge: be open about how the differentiation is implemented, ensure that you're not discriminating against sensitive groups, and consider the impact on the privacy of the data and model subjects. An intriguing related mechanism is that of behaviour modification, as discussed by Shmueli [397], where a person's behaviour might be nudged towards the predictions, so as to make the predictions true.

The DeepFake technology, based on deep learning, has led to numerous fake videos, often funny and sometimes even used in the professional entertainment industry, but unfortunately also in scams and fake porn videos. Solutions include the use of DeepFake detection models and simply educating the public about the existence and dangers of DeepFake.

How to set up data science ethics in a company is discussed in the Governance section. The three discussed guidelines included: setting up an ethical oversight committee, establishing a policy, and then implementing this. Finally, the unintended consequences of data science models can go far beyond the immediate impact on the model subject. This ranged from a bot that was designed with good intentions, yet was tricked into racist and sexists comments, to the potential social dramas that might occur as data science and the accompanying automation will have a drastic impact on jobs as we know them.

7

Conclusion

Data science has had a tremendous positive effect on companies and persons, and is quickly becoming a standard practice at all but a few companies. The potential negative effects, however, can be severe, unexpected, and come with a large cost. Those working on the frontier of data science technology are bound to be confronted with ethical issues, as was illustrated with the many cautionary tales. Good intentions alone are not enough. Being ethical requires thorough ethical thinking. First, in understanding the trade-off between utility of the data and the importance of the ethical aspects. What is the right balance, and what is considered ethical, depends on the context: what data is used, how is it gathered and preprocessed, what application and modelling is foreseen? Have an open conversation on where you want to be, and report on this chosen *golden mean* (dixit Aristotle) transparently. This data science ethics equilibrium will then determine which of the discussed techniques are suitable to be implemented.

To conclude, let's revisit what data science ethics entails:

- **Data science ethics requires an open discussion.** Don't shy away from openly discussing the utility of data. To avoid a fear of saying the wrong thing, specific points of view can be assigned in the discussion: some need to argue in favour of the utility of data, others in favour of privacy and fairness, irrespective of what their actual opinion is. These discussions should lead to a consensus on the chosen equilibrium. A summary of the discussion can be reported on transparently. Obviously, there might be a different report to be provided to the different stakeholders, as company secrets or technical details are not suitable to be disclosed to all. To train yourself in having such open, constructive discussions, the 10 Discussions from this book can serve as cases to start off with. Similarly, don't shy away from including an ethical section in your next data science report, presentation, or email, which may spur a new discussion on the topic. On a company level, future non-financial reports and standards might include a data science ethics section as well.

Data Science Ethics. David Martens, Oxford University Press.
© David Martens (2022). DOI: 10.1093/oso/9780192847263.003.0007

- **Data science ethics is not easy.** It requires training, effort, and time. Not only to have the open discussions, but also to understand the potential issues and concepts, and grasp the existing techniques that might be of relevance. As both the data science and data science ethics research domains are quickly evolving, keeping up with good practices will require an investment in time and resources. The commitment from senior management to invest in data science ethics and acknowledgment of its importance is therefore likely a necessary condition. Let it also be a call to be mild. Working on the edge of data science will unavoidably lead to ethical issues that were previously not widely acknowledged or even known off. The possibility to learn from these cautionary tales can make you a wiser and better data scientist, manager, or even citizen. Similarly, be mild as you judge and discuss historical cases and cautionary tales through the lens of currently obtained insights. Ethics is dynamic, and changes over time, location, and application. Beware that you too might unintentionally become the topic of future cautionary tales.

- **Data science ethics does *not* require you to not use any data at all.** Being ethical does not ask you to do nothing with data. Aristotle showed us the way here, and motivated that deficiency in itself is unethical. Rather, one should find the right balance between deficiency and excess. Taking a point of view that only focuses on privacy or fairness can be valid to protect the rights of some stakeholder, like the data subject. But these arguments should be included in a bigger discussion on finding the right balance between deficiency and excess.

- **Data science ethics is more than a checklist.** Due to the dynamic and subjective nature of what ethics is, general checklists are bound to be irrelevant to some contexts. Similarly, as the field is continuously evolving, checklists have a short lifespan and will need to be updated frequently. They can of course be valuable in some situations, for example to implement a company's ethical policy, but then require parallel training and transparent reporting into which concepts and principles lie behind the list, and procedures to keep the list update. The FAT Flow framework that guided the chapters of this book can serve as one way to structure your data science ethics.

- **Data science ethics can bring value.** Being ethical in your data science will avoid costs: reputation cost and the potential legal and financial costs that come with it. But there is also value to be gained: we discussed for example how explaining prediction models or individual predictions can reveal data quality issues or potential model bias. Choosing the right privacy-friendly data gathering mechanism might lead to data from a large sample to become available, which then might lead to better data science models. Similarly, choosing a suitable privacy-preserving data mining setup can also lead to better models. For example, federated learning could help to leverage data from more data sources. These improvements to the data science model can lead to economic advantages, be it more efficiency, less costs, more profit, better customer service, etc. As citizens these days increasingly expect companies to behave ethically, ethical data science can even be a marketing instrument to attract or up-sell to these consumers. So make data science ethics a positive thing, not a necessary evil that you have to comply with so as to avoid costs. Make it fun, through constructive, interesting discussions, by learning about new technologies, investigating how data science ethics concepts and techniques can be leveraged in your institution, and telling the story of how data science ethics has improved the life of your customers, employees, or society as a whole.

The future of data science ethics is bright. Data has taken up an immensely important role in our society, and this cannot be turned back. Although we are often confronted with reported data leaks, privacy concerns and discrimination controversies, the global attention in data science seems to be shifting in the right direction. In my humble opinion, the culture and technology to deal with the ethical implications are, and will be, changing for the better. Simply the fact that you have read this book is a positive indication in that direction.

Bibliography

[1] Abalone (22 June 2014). Lessons from NYC's improperly anonymized taxi logs. https://news.ycombinator.com/item?id=7927034. Hacker News, Online, accessed 1 May 2020.

[2] Acar, Abbas, Aksu, Hidayet, Uluagac, A. Selcuk, and Conti, Mauro (2018). A survey on homomorphic encryption schemes: Theory and implementation. *ACM Computing Surveys*, **51**(4), 508–516.

[3] Acemoglu, Daron and Robinson, James A. (2012). *Why Nations Fail: The Origins of Power, Prosperity and Poverty* (1st edn). Crown, New York.

[4] Aggarwal, Charu C. (2005). On k-anonymity and the curse of dimensionality. In *Proceedings of the 31st International Conference on Very Large Data Bases*, VLDB '05, pp. 901–909. VLDB Endowment.

[5] Agrawal, Aditi (12 November 2019). New York regulator orders probe into Goldman Sachs' credit card practices over Apple card and sexism. https://www.medianama.com/2019/11/223-apple-card-sexism-goldman-sachs/. Medianama, Online, accessed 21 January 2021.

[6] Alba, Davey (16 July 2019). A Google VP told the US Senate the company has "terminated" the Chinese search app dragonfly. https://www.buzzfeednews.com/article/daveyalba/google-project-dragonfly-terminated-senate-hearing. Online, accessed 25 May 2021.

[7] Alcine, Jacky (June 29, 2015). Twitter post on Google Photos app. https://twitter.com/jackyalcine/status/615329515909156865. Twitter, Online, accessed May 1, 2020.

[8] Allea (2017). The European Code of Conduct for Research Integrity. https://www.allea.org/wp-content/uploads/2017/05/ALLEA-European-Code-of-Conduct-for-Research-Integrity-2017.pdf. Online, accessed 1 October 2020.

[9] Amatriain, Xavier and Basilico, Justin (6 April 2012). Netflix recommendations: Beyond the 5 stars. https://netflixtechblog.com/netflix-recommendations-beyond-the-5-stars-part-1-55838468f429. Netflix Technology Blog, Online, accessed 1 May 2020.

[10] Anderson, Alan Ross (1951). A note on subjunctive and counterfactual conditionals. *Analysis*, **12**(2), 35–38.

[11] Andrews, Robert, Diederich, Joachim, and Tickle, Alan B. (1995). Survey and critique of techniques for extracting rules from trained artificial neural networks. *Knowledge-Based Systems*, **8**(6), 373–389.

[12] Angwin, Julia, Larson, Jeff, Mattu, Surya, and Kirchne, Lauren (23 May 2016). Machine bias. https://www.propublica.org/article/machine-bias-risk-assessments-in-criminal-sentencing. ProPublica, Online, accessed 21 January 2021.

[13] Angwin, Julia, Larson, Jeff, Mattu, Surya, and Kirchner, Lauren (2016, May). Machine bias. https://medium.com/the-official-integrate-ai-blog/towards-fair-machine-learning-models-acfc6c4f9ee5. Online; accessed 22-03-2019.

[14] Anthony, Lasse F. Wolff, Kanding, Benjamin, and Selvan, Raghavendra (2020). Carbontracker: Tracking and predicting the carbon footprint of training deep learning models arXiv:2007.03051 [cs.CY].

[15] Apple (2020). Apple differential privacy technical overview. https://www.apple.com/privacy/docs/Differential_Privacy_Overview.pdf. Apple, Online, accessed 1 October 2020.

[16] Aral, S. (2020). *The Hype Machine*. HarperCollins Publishers.

[17] Aristotle (2000). *Aristotle: Nicomachean Ethics*. Cambridge Texts in the History of Philosophy. Cambridge University Press.

[18] Arrieta, Alejandro Barredo, Diaz-Rodriguez, Natalia, Ser, Javier Del, Bennetot, Adrien, Tabik, Siham, Barbado, Alberto, Garcia, Salvador, Gil-Lopez, Sergio, Molina, Daniel, Benjamins, Richard, Chatila, Raja, and Herrera, Francisco (2019). Explainable artificial intelligence (XAI): Concepts, taxonomies, opportunities and challenges toward responsible AI.

[19] Article 29 Data Protection Working Party (2010). Opinion 3/2010 on the principle of accountability. Technical Report WP173 00062/10/EN, European Commission.

[20] Article 29 Data Protection Working Party (2014). Opinion 06/2014 on the notion of legitimate interests of the data controller. Technical Report WP217 844/14/EN, European Commission.

[21] Article 29 Data Protection Working Party (2014, April). Opinion 06/2014 on the notion of legitimate interests of the data controller under article 7 of directive 95/46/ec. Technical Report 844/14/EN, European Commission.

[22] Article 29 Data Protection Working Party (2017). Guidelines on automated individual decision-making and profiling for the purposes of regulation 2016/679. Technical Report WP 251 17/EN, European Commission.

[23] Article 29 Data Protection Working Party (2017). Guidelines on automated individual decision-making and profiling for the purposes of regulation 2016/679. Technical Report WP173 00062/10/EN, European Commission.

[24] Aschwanden, Christie (24 November 2015). Not even scientists can easily explain p-values. https://fivethirtyeight.com/features/not-even-scientists-can-easily-explain-p-values/. Online, accessed 25 May 2021.

[25] Austin, Algernon, Bucknor, Cherrie, Cashman, Kevin, and Rockeymoore, Maya (2017). Stick shift: Autonomous vehicles, driving jobs, and the future of work. https://globalpolicysolutions.org/wp-content/uploads/2017/03/Stick-Shift-Autonomous-Vehicles.pdf. Center for Global Policy Solutions, Online, accessed 8 March 2021.

[26] Autor, David, Mindell, Dvid, and Reynolds, Elisabeth (2020). *The Work of the Future: Building Better Jobs in an Age of Intelligent machines*. https://workofthefuture.mit.edu/research-post/the-work-of-the-future-building-better-jobs-in-an-age-of-intelligent-machines/. MIT.

[27] Autor, David H. and Dorn, David (2013, August). The growth of low-skill service jobs and the polarization of the us labor market. *American Economic Review*, **103**(5), 1553–1597.

[28] Awad, Edmond, Dsouza, Sohan, Kim, Richard, Schulz, Jonathan, Henrich, Joseph, Shariff, Azim, Bonnefon, Jean-François, and Rahwan, Iyad (2018, 11). The moral machine experiment. *Nature*, **563**.

[29] Baesens, Bart (2014). *Analytics in a Big Data World: The Essential Guide to Data Science and Its Applications* (1st edn). Wiley Publishing.

[30] Bahajji, Zineb Ait and Illyes, Gary (6 August 2014). HTTPS as a ranking signal. https://webmasters.googleblog.com/2014/08/https-as-ranking-signal.html. Google WebMaster Central Blog, Online, accessed 19 July 2019.

[31] Bake, Monya (2016, May). 1,500 scientists lift the lid on reproducibility. *Nature*, **533**, 452–454.

[32] Bakos, Yannis, Marotta-Wurgler, Florencia, and Trossen, David R. (2009, August). Does Anyone Read the Fine Print? Testing a Law and Economics Approach to Standard Form Contracts. Working Papers 09-04, NET Institute.

[33] Balazs, Kovacs Zoltan (15 April 2019). New Hungarian data protection fines. https://eugdpr.blog.hu/2019/04/15/new_hungarian_data_protection_fines. EU GDPR Blog, Online, accessed 7 March 2019.

[34] Ball, James, Borger, Julian, and Greenwald, Glenn (6 September 2013). Revealed: how US and UK spy agencies defeat internet privacy and security. https://www.theguardian.com/world/2013/sep/05/nsa-gchq-encryption-codes-security. Online, accessed 25 May 2021.

[35] Barbaro, Michael and Jr., Tom Zeller (9 August 2006). Web searchers' identities traced on aol. https://www.nytimes.com/2006/08/09/technology/08cnd-aol.html. New York Times, Online, accessed 1 May 2020.

[36] Barbera, Pablo, Jost, John T., Nagler, Jonathan, Tucker, Joshua A., and Bonneau, Richard (2015). Tweeting from left to right: Is online political communication more than an echo chamber? *Psychological Science*, **26**(10), 1531–1542.

[37] Barkalow, Carol (1990). *In the Men's House*. Poseidon Press.

[38] Barocas, Solon and Nissenbaum, Helen (2014). *Big Data's End Run around Anonymity and Consent*, pp. 44–75. Cambridge University Press.

[39] Barocas, Solon and Selbst, Andrew D. (2016). Big Data's disparate impact. *California Law Review*, **671**.

[40] Barocas, Solon, Selbst, Andrew D., and Raghavan, Manish (2020). The hidden assumptions behind counterfactual explanations and principal reasons. In *Proceedings of the 2020 Conference on Fairness, Accountability, and Transparency*, FAT* '20, New York, NY, USA, pp. 80–89. Association for Computing Machinery.

[41] Barr, Alistair (1 July 2015). Google mistakenly tags black people as 'gorillas,' showing limits of algorithms. https://blogs.wsj.com/digits/2015/07/01/google-mistakenly-tags-black-people-as-gorillas-showing-limits-of-algorithms/. The Wall Street Journal, Online, accessed 1 May 2020.

[42] Bastos, Marcos and Walker, Shawn T. (11 April 2018). Facebook's data lockdown is a disaster for academic researchers. https://theconversation.com/facebooks-data-lockdown-is-a-disaster-for-academic-researchers-94533. The Conversation, Online, accessed 1 May 2020.

[43] BBC (25 May 2018). GDPR: US news sites unavailable to EU users under new rules. https://www.bbc.com/news/world-europe-44248448. BBC Online, accessed 1 October 2020.

[44] BBC News (23 March 2010). China condemns decision by Google to lift censorship. http://news.bbc.co.uk/2/hi/asia-pacific/8582233.stm. BBC News Online, accessed 1 October 2020.

[45] BBC News (5 May 2021). Facebook's Trump ban upheld by oversight board for now. https://www.bbc.com/news/technology-56985583. Online, accessed 25 May 2021.

[46] BBC News—Magazine Monitor (29 July 2014). Okcupid: Who believes compatibility ratings on dating websites? https://www.bbc.com/news/blogs-magazine-monitor-28543248. Online, accessed 25 May 2021.

[47] Beecher, Henry K. (1966). Ethics and clinical research. *New England Journal of Medicine*, **274**(24), 1354–1360.

[48] Begley, C. Glenn and Ellis, Lee M. (2012, March). Drug development: Raise standards for preclinical cancer research. *Nature*, **483**(7391), 531–533.

[49] Begley, Sharon (28 March 2012). In cancer science, many 'discoveries' don't hold up. https://www.reuters.com/article/us-science-cancer-idUSBRE82R12P20120328. Reuters, Online, accessed 1 October 2020.

[50] Belongia, Edward A. and Naleway, Allison L. (2003). Smallpox vaccine: the good, the bad, and the ugly. *Clinical Medicine & Research*, **1**(2), 87–92.

[51] Ben-Sasson, E., Chiesa, A., Garman, C., Green, M., Miers, I., Tromer, E., and Virza, M. (2014). Zerocash: Decentralized anonymous payments from bitcoin. In *2014 IEEE Symposium on Security and Privacy*, pp. 459–474.

[52] Benbunan-Fich, Raquel (2017). The ethics of online research with unsuspecting users: From a/b testing to c/d experimentation. *Research Ethics*, **13**(3-4), 200–218.

[53] Benner, Katie and Lichtblau, Eric (28 March 2016). U.S. says it has unlocked iPhone without Apple. https://www.nytimes.com/2016/03/29/technology/apple-iphone-fbi-justice-department-case.html. Online, accessed 25 May 2021.

[54] Bershidsky, Leonid (7 July 2014). Larry Page's slacker utopia. https://www.bloomberg.com/opinion/articles/2014-07-07/larry-page-s-slacker-utopia. Online, accessed 8 March 2021.

[55] Bestavros, Azer, Lapets, Andrei, and Varia, Mayank (2017, January). User-centric distributed solutions for privacy-preserving analytics. *Commun. ACM*, **60**(2), 37–39.

[56] Bhattacharjee, Yudhijit (28 April 2013). The mind of a con man. https://www.nytimes.com/2013/04/28/magazine/diederik-stapels-audacious-academic-fraud.html. New York Times, Online, accessed 1 October 2020.

[57] Blaze, Matt (15 December 2015). A key under the doormat isn't safe. neither is an encryption backdoor. https://www.washingtonpost.com/news/in-theory/wp/2015/12/15/how-the-nsa-tried-to-build-safe-encryption-but-failed/. The Washington Post, Online, accessed 1 May 2020.

[58] Bogetoft, Peter, Christensen, Dan Lund, Damgård, Ivan, Geisler, Martin, Jakobsen, Thomas, Krøigaard, Mikkel, Nielsen, Janus Dam, Nielsen, Jesper Buus, Nielsen, Kurt, Pagter, Jakob, Schwartzbach, Michael, and Toft, Tomas (2009). *Secure Multiparty Computation Goes Live*, pp. 325–343. Springer-Verlag, Berlin, Heidelberg.

[59] Bolukbasi, Tolga, Chang, Kai-Wei, Zou, James, Saligrama, Venkatesh, and Kalai, Adam (2016). Man is to computer programmer as woman is to homemaker? debiasing word embeddings. In *Proceedings of the 30th International Conference on Neural Information Processing Systems*, NIPS'16, Red Hook, NY, USA, pp. 4356–4364. Curran Associates Inc.

[60] Bossmann, Julia (21 October 2016). Top 9 ethical issues in artificial intelligence. https://www.weforum.org/agenda/2016/10/top-10-ethical-issues-in-artificial-intelligence/. World Economic Forum, Online, accessed 25 May 2021.

[61] Bovens, Mark (2006, 01). Analysing and assessing public accountability. a conceptual framework. *CONNEX and EUROGOV Networks, European Governance Papers (EUROGOV)*.

[62] Bradlow, Eric, Gangwar, Manish, Kopalle, Praveen, and Voleti, Sudhir (2017, 03). The role of big data and predictive analytics in retailing. *Journal of Retailing*, **93**, 79–95.

[63] Brandt, Allan M. (1978). Racism and research: The case of the tuskegee syphilis study. *Hastings Center Report*, **8**(6), 21–29.

[64] Branwen, Gwern (14 August 2019). The neural net tank urban legend. https://www.gwern.net/Tanks. gwern.net. Online, accessed 8 March 2021.

[65] Breiman, Leo (2001). Random forests. *Machine Learning*, **45**(1), 5–32.

[66] Brewster, Thomas (22 January 2017). Forget about backdoors, this is the data whatsapp actually hands to cops. https://www.forbes.com/sites/thomasbrewster/2017/01/22/whatsapp-facebook-backdoor-government-data-request/#34aa039b1030. Forbes, Online, accessed 28 February 2020.

[67] Brey, Philip (2004, 05). Ethical aspects of facial recognition systems in public places. *Journal of Information, Communication and Ethics in Society*, **2**, 97–109.

[68] Brian Fung (28 July 2014). Okcupid reveals it's been lying to some of its users. just to see what'll happen. https://www.washingtonpost.com/news/the-switch/wp/2014/07/28/okcupid-reveals-its-been-lying-to-some-of-its-users-just-to-see-whatll-happen/. Washington Post, Online, accessed 28 October 2019.

[69] Brodkin, Jon (8 December 2009). Facebook halts beacon, gives $9.5m to settle lawsuit. https://www.pcworld.com/article/184029/facebook_halts_beacon_gives_9_5_million_to_settle_lawsuit.html. PC World, Online, accessed 1 June 2020.

[70] Brown, Tom B., Mann, Benjamin, Ryder, Nick, Subbiah, Melanie, Kaplan, Jared, Dhariwal, Prafulla, Neelakantan, Arvind, Shyam, Pranav, Sastry, Girish, Askell, Amanda, Agarwal, Sandhini, Herbert-Voss, Ariel, Krueger, Gretchen, Henighan, Tom, Child, Rewon, Ramesh, Aditya, Ziegler, Daniel M., Wu, Jeffrey, Winter, Clemens, Hesse, Christopher, Chen, Mark, Sigler, Eric, Litwin, Mateusz, Gray, Scott, Chess, Benjamin, Clark, Jack, Berner, Christopher, McCandlish, Sam, Radford, Alec, Sutskever, Ilya, and Amodei, Dario (2020). Language models are few-shot learners. *CoRR*, abs/2005.14165.

[71] Brunton, Finn and Nissenbaum, Helen (2015). *Obfuscation: A User's Guide for Privacy and Protest*. The MIT Press.

[72] Buolamwini, Joy and Gebru, Timnit (2018, 23–24 February). Gender shades: Intersectional accuracy disparities in commercial gender classification. In *Proceedings of the 1st Conference on Fairness, Accountability and Transparency* (ed. S. A. Friedler and C. Wilson), Volume 81 of *Proceedings of Machine Learning Research*, New York, NY, USA, pp. 77–91. PMLR.

[73] Buolamwini, Joy and Gebru, Timnit (2018, 23–24 Feb). Gender shades: Intersectional accuracy disparities in commercial gender classification. Volume 81 of *Proceedings of Machine Learning Research*, New York, NY, USA, pp. 77–91. PMLR.

[74] Bureau of Export Administration (2000, January). Revised U.S. encryption export control regulations. https://epic.org/crypto/export_controls/regs_1_00.html. Electronic Privacy Information Center, Online, accessed 13 September 2019.

[75] BuzzFeedVideo (17 April 2018). You won't believe what Obama says in this video! ;). https://www.youtube.com/watch?v=cQ54GDm1eL0. Online, accessed 8 March 2021.

[76] Calders, Toon and Verwer, Sicco (2010, September). Three naive Bayes approaches for discrimination-free classification. *Data Min. Knowl. Discov.*, **21**(2), 277–292.

[77] Camenisch, Jan, Chaabouni, Rafik, and Shelat, Abhi (2008). Efficient protocols for set membership and range proofs. In *Proceedings of the 14th International Conference on the Theory and Application of Cryptology and Information Security: Advances in Cryptology*, ASIACRYPT '08, Berlin, Heidelberg, pp. 234–252. Springer-Verlag.

[78] Canavan, John E. (2000). *Fundamentals of Network Security*. Artech House.

[79] Candela, Joaquin Quinonero (20 January 2019). Facebook and the Technical University of Munich announce new independent TUM Institute for Ethics

in Artificial Intelligence. https://about.fb.com/news/2019/01/tum-institute-for-ethics-in-ai/. Online, acces-sed 25 May 2021.

[80] CBS News (5 February 2020). Google, Youtube, Venmo and Linkedin send cease-and-desist letters to facial recognition app that helps law enforcement. https://www.cbsnews.com/news/clearview-ai-google-youtube-send-cease-and-desist-letter-to-facial-recognition-app/. CBS News, Online, accessed 1 October 2020.

[81] Cerullo, Megan (July 10, 2020). Universal basic income cause gets a $3 million boost from Jack Dorsey. https://www.cbsnews.com/news/jack-dorsey-twitter-guaranteed-income-programs-3-million-donation/. Online, accessed 8 March 2021.

[82] Chalkias, Kostas (13 September 2017). Demonstrate how zero-knowledge proofs work without using maths. https://www.linkedin.com/pulse/demonstrate-how-zero-knowledge-proofs-work-without-using-chalkias/. Online, accessed 25 May 2021.

[83] Chambers, Chris (1 July 2014). Facebook fiasco: was Cornell's study of 'emotional contagion' an ethics breach? https://www.theguardian.com/science/head-quarters/2014/jul/01/facebook-cornell-study-emotional-contagion-ethics-breach. Online, accessed 25 May 2021.

[84] Chan, Siu On (2016). Advanced topic: Zero-knowledge proofs. https://www.cse.cuhk.edu.hk/~siuon/csci3130-f16/slides/lec24.pdf. Online, accessed 25 May 2021.

[85] Chappell, Andrew (29 April 2019). Danish DPA's recent fine offers insight on GDPR Article 5 enforcement. https://iapp.org/news/a/danish-dpas-recent-fine-offers-guidance-on-article-5/. IAPP, Online, accessed 7 March 2019.

[86] Chen, Daizhuo, Fraiberger, Samuel P., Moakler, Robert, and Provost, Foster J. (2017). Enhancing transparency and control when drawing data-driven inferences about individuals. *Big Data*, **5**(3), 197–212.

[87] Chen, Hao, Laine, Kim, and Rindal, Peter (2017, October). Fast private set intersection from homomorphic encryption. In *CCS '17 Proceedings of the 2017 ACM SIGSAC Conference on Computer and Communications Security* (CCS '17 Proceedings of the 2017 ACM SIGSAC Conference on Computer and Communications Security edn), pp. 1243–1255. ACM New York, NY, USA.

[88] Chen, Yen-Liang, Tang, Kwei, Shen, Ren-Jie, and Hu, Ya-Han (2005, August). Market basket analysis in a multiple store environment. *Decis. Support Syst.*, **40**(2), 339–354.

[89] Chipman, Jason, Ferraro, Matthew, and Preston, Stephen (24 December 2019). First federal legislation on deepfakes signed into law. https://www.jdsupra.com/legalnews/first-federal-legislation-on-deepfakes-42346/. Online, accessed 8 March 2021.

[90] Chouldechova, Alexandra (2016, 10). Fair prediction with disparate impact: A study of bias in recidivism prediction instruments. *Big Data*, **5**.

[91] Christoper Mims (4 June 2019). The day when computers can break all encryption is coming. https://www.wsj.com/articles/the-race-to-save-encryption-11559646737. The Wall Street Journal, Online, accessed 1 August 2019.

[92] Cinelli, Matteo, De Francisci Morales, Gianmarco, Galeazzi, Alessandro, Quattrociocchi, Walter, and Starnini, Michele (2021). The echo chamber effect on social media. *Proceedings of the National Academy of Sciences*, **118**(9).

[93] Clifford, Catherine (18 June 2018). Elon Musk: Free cash handouts 'will be necessary' if robots take humans' jobs. https://www.cnbc.com/2018/06/18/elon-musk-automated-jobs-could-make-ubi-cash-handouts-necessary.html. Online, accessed 8 March 2021.

[94] CNN (January 26, 2006). Google to censor itself in China. http://edition.cnn.com/2006/BUSINESS/01/25/google.china/. CNN.com, Online, accessed 1 October 2020.

[95] Constantin, Andres (2018, 12). Human subject research. *Health and Human Rights Journal*, **20**(2), 137–148.

[96] COPE (2018). Editor and reviewers requiring authors to cite their own work. https://publicationethics.org/case/editor-and-reviewers-requiring-authors-cite-their-own-work. Online, accessed 25 May 2021.

[97] Craven, Mark W. and Shavlik, Jude W. (1995). Extracting tree-structured representations of trained networks. In *Proceedings of the 8th International Conference on Neural Information Processing Systems*, NIPS'95, Cambridge, MA, USA, pp. 24–30. MIT Press.

[98] Crawford, Kate (10 May 2013). Think again. https://foreignpolicy.com/2013/05/10/think-again-big-data/. Foreign policyt, Online, accessed 11 November 2019.

[99] Crawford, Kate and Schultz, Jason (2014). Big data and due process: Toward a framework to redress predictive privacy harms. *Boston College Law Review*, **55**(1).

[100] Crook, Jonathan and Banasik, John (2004, 04). Does reject inference really improve the performance of application scoring models? *Journal of Banking & Finance*, **28**, 857–874.

[101] Dalessandro, Brian, Hook, Rod, Perlich, Claudia, and Provost, Foster (2012, 10). Evaluating and optimizing online advertising: Forget the click, but there are good proxies. *Big Data*, **3**.

[102] Dastin, Jeffrey (10 October 2018). Amazon scraps secret AI recruiting tool that showed bias against women. https://www.reuters.com/article/us-amazon-com-jobs-automation-insight/amazon-scraps-secret-ai-recruiting-tool-that-showed-bias-against-women-idUSKCN1MK08G. Reuters, Online, accessed 28 February 2020.

[103] Davenport, Thomas H. and Patil, D. J. (2012, October). Data scientist: The sexiest job of the 21st century. *Harvard Business Review*.

[104] De Cnudde, Sofie, Moeyersoms, Julie, Stankova, Marija, Tobback, Ellen, Javaly, Vinayak, and Martens, David (2019). What does your Facebook profile reveal about your creditworthiness? using alternative data for microfinance. *Journal of the Operational Research Society* (3).

[105] de Fortuny, Enric Junqué and Martens, David (2015). Active learning-based pedagogical rule extraction. *IEEE Trans. Neural Networks Learn. Syst.*, **26**(11), 2664–2677.

[106] de Montjoye, Yves-Alexandre, Hidalgo, Cesar, Verleysen, Michel, and Blondel, Vincent (2013, 03). Unique in the crowd: The privacy bounds of human mobility. *Scientific reports*, **3**, 1376.

[107] Dean Takahashi (14 July 2019). IBM research explains how quantum computing works and why it matters. https://venturebeat.com/2019/07/14/ibm-research-explains-how-quantum-computing-works-and-could-be-the-the-supercomputer-of-the-future/. Venture Beat, Online, accessed 4 August 2019.

[108] Dearden, Lizzie (7 May 2019). Facial recognition wrongly identifies public as potential criminals 96% of time, figures reveal. https://www.independent.co.uk/news/uk/home-news/facial-recognition-london-inaccurate-met-police-trials-a8898946.html. Independent, Online, accessed 1 June 2020.

[109] Dedman, Bill (1 May 1988). Atlanta blacks losing in home loans scramble. http://powerreporting.com/color/color_of_money.pdf. Online, accessed 21 January 2021.

[110] Dedman, Bill (May 2, 1988). Southside treated like banks' stepchild? http://powerreporting.com/color/color_of_money.pdf. Online, accessed 21 January 2021.

[111] Delfs, Hans and Knebl, Helmut (2002). *Introduction to Cryptography: Principles and Applications*. Springer-Verlag, Berlin, Heidelberg.

[112] Devjyot Ghoshal (28 March 2018). Mapped: The breathtaking global reach of cambridge analytica's parent company. https://qz.com/1239762/cambridge-analytica-scandal-all-the-countries-where-scl-elections-claims-to-have-worked/. Quartz, Online, accessed 28 October 2019.

[113] Diamond, Jeremy (20 February 2016). Trump calls for Apple boycott. https://edition.cnn.com/2016/02/19/politics/donald-trump-apple-boycott/index.html. Online, accessed 25 May 2021.

[114] Dickinson, Charles (1859). *A Tale of Two Cities*. Chapman and Hall.

[115] Domingos, Pedro (1999). The role of Occam's Razor in knowledge discovery. *Data Mining and Knowledge Discovery*, **3**, 409–425.

[116] Don Gotterbarn et al. (2018). ACM code of ethics and professional conduct. Online, accessed 8 March 2021.

[117] Drummond, David (12 January 2010). A new approach to China. https://googleblog.blogspot.com/2010/01/new-approach-to-china.html. Google Official Blog, Online, accessed 1 October 2020.

[118] Duhigg, Charles (16 Febuary 2012). How companies learn your secrets. https://www.nytimes.com/2012/02/19/magazine/shopping-habits.html. New York Times, Online, accessed 1 June 2020.

[119] Duman, Sevtap, Kalai, Adam Tauman, Leiserson, Max, Mackey, Lester, and Sursesh, Harini (2017). Gender bias in word embeddings. http://wordbias.umiacs.umd.edu/. This was a result of the Hacking Discrimination hackathon held at Microsoft New England Research & Development (NERD) center on 2 February 2017, Online, accessed 12 January 2021.

[120] DutchNews (8 September 2011). Meat eaters absolved, professor in the dock. https://www.dutchnews.nl/features/2011/09/meat_eaters_absolved_professor/. Online, accessed 1 October 2020.

[121] Dwork, Cynthia, Hardt, Moritz, Pitassi, Toniann, Reingold, Omer, and Zemel, Richard (2012). Fairness through awareness. In *Proceedings of the 3rd Innovations in Theoretical Computer Science Conference*, ITCS '12, New York, NY, USA, pp. 214–226. Association for Computing Machinery.

[122] Dwork, Cynthia, McSherry, Frank, Nissim, Kobbi, and Smith, Adam (2006). Calibrating noise to sensitivity in private data analysis. In *Theory of Cryptography* (ed. S. Halevi and T. Rabin), Berlin, Heidelberg, pp. 265–284. Springer Berlin Heidelberg.

[123] Dwork, Cynthia and Roth, Aaron (2014). The algorithmic foundations of differential privacy. *Foundations and Trends in Theoretical Computer Science*, **9**(3-4), 211–407.

[124] Edmonds, David (2014). *Would You Kill the Fat Man?: The Trolley Problem and What Your Answer Tells Us about Right and Wrong*. Princeton University Press.

[125] Elizabeth Vowell (26 April 2018). Three years later, investigators still search for pregnant mother's killer. https://www.wafb.com/story/38045081/three-years-later-investigators-still-search-for-pregnant-mothers-killer/. WAFB 9, Online, accessed 13 September 2019.

[126] Ellenberg, Jordan (2014). *How Not to Be Wrong: The Power of Mathematical Thinking.* The Penguin Press.

[127] Ellis, James (2010). The history of non-secret encryption. *Cryptologia,* **23**(3), 267–273.

[128] Erlingsson, Úlfar, Pihur, Vasyl, and Korolova, Aleksandra (2014). Rappor: Randomized aggregatable privacy-preserving ordinal response. In *Proceedings of the 2014 ACM SIGSAC Conference on Computer and Communications Security,* CCS '14, New York, NY, USA, pp. 1054–1067. Association for Computing Machinery.

[129] Esteva, Andre, Kuprel, Brett, Novoa, Roberto A., Ko, Justin, Swetter, Susan M., Blau, Helen M., and Thrun, Sebastian (2017, January). Dermatologist-level classification of skin cancer with deep neural networks. *Nature,* **542**, 115–118.

[130] Ethics Commission Automated and Connected Driving (2017). Ethics Commission Automated and Connected Driving.

[131] European Commission (21 April 2021). Europe fit for the digital age: Commission proposes new rules and actions for excellence and trust in artificial intelligence. https://ec.europa.eu/commission/presscorner/detail/en/ip_21_1682. Online, accessed 25 May 2021.

[132] European Court of Human Rights (2010, December). Case of Roman Xakharov v. Russia. Technical Report Application no. 47143/06, European Court of Human Rights.

[133] European Data Protection Board (25 March 2019). The Danish Data Protection Agency proposes a dkk 1,2 million fine for Danish taxi company. https://edpb.europa.eu/news/national-news/2019/danish-data-protection-agency-proposes-dkk-12-million-fine-danish-taxi_en. Online, accessed 25 May 2021.

[134] European Data Protection Board (21 May 2019). First significant fine was imposed for the breaches of the general data protection regulation in lithuania. https://edpb.europa.eu/news/national-news/2019/first-significant-fine-was-imposed-breaches-general-data-protection_lv. Online; accessed 7 March 2019.

[135] Evan Selinger (11 October 2018). Stop saying privacy is dead. https://medium.com/s/story/stop-saying-privacy-is-dead-513dda573071. Medium, Online, accessed 13 September 2019.

[136] Evans, David, Kolesnikov, Vladimir, and Rosulek, Mike (2018). A pragmatic introduction to secure multi-party computation. *Foundations and Trends in Privacy and Security,* **2**(2-3), 70–246.

[137] Facebook (31 March 2019). Why am I seeing this? we have an answer for you. https://about.fb.com/news/2019/03/why-am-i-seeing-this/. Online, accessed 25 May 2021.

[138] Facebook AI (31 March 2021). How we're using fairness flow to help build ai that works better for everyone. https://ai.facebook.com/blog/how-were-using-fairness-flow-to-help-build-ai-that-works-better-for-everyone/. Online, accessed 25 May 2021.

[139] Fazzini, Kate and Kolodny, Lora (29 March 2019). Tesla cars keep more data than you think, including this video of a crash that totaled a model 3. https://www.cnbc.com/2019/03/29/tesla-model-3-keeps-data-like-crash-videos-location-phone-contacts.html. CNBC, Online, accessed 19 July 2019.

[140] Federal Committee on Statistical Methodology (2005). Report on statistical disclosure limitation methodology. Technical Report Statistical Policy Working Papaer 22, Federal Committee on Statistical Methodology.

[141] Federal Trade Commission (July 24, 2019). FTC imposes $5 billion penalty and sweeping new privacy restrictions on Facebook. https://www.ftc.gov/news-

events/press-releases/2019/07/ftc-imposes-5-billion-penalty-sweeping-new-privacy-restrictions. Online, accessed 25 May 2021.

[142] Fernandez-Loria, Carlos, Provost, Foster, and Han, Xintian (2020). Explaining data-driven decisions made by ai systems: The counterfactual approach. arXiv:2001.07417 [cs.LG].

[143] Fifield, Anna (29 August 2019). Jack Ma, once proponent of 12-hour work-days, now foresees 12-hour workweeks. https://www.washingtonpost.com/world/asia_pacific/jack-ma-proponent-of-12-hour-work-days-foresees-12-hour-workweeks/2019/08/29/fd081370-ca2a-11e9-9615-8f1a32962e04_story.html. Online, accessed 8 March 2021.

[144] Finley, Mary Lou (8 February 2016). The Chicago Freedom Movement and the fight for fair lending. https://www.chicagoreporter.com/the-chicago-freedom-movement-and-the-fight-for-fair-lending/. Online, accessed 21 January 2021.

[145] Five Country Ministerial (2018). Statement of principles on access to evidence and encryption. https://www.ag.gov.au/About/CommitteesandCouncils/Documents/joint-statement-principles-access-evidence.pdf. Australian Government, Attorney-General's Department, Online, accessed 13 September 2019.

[146] Fortunato, Santo, Bergstrom, Carl T., Börner, Katy, Evans, James A., Helbing, Dirk, Milojević, Staša, Petersen, Alexander M., Radicchi, Filippo, Sinatra, Roberta, Uzzi, Brian, Vespignani, Alessandro, Waltman, Ludo, Wang, Dashun, and Barabási, Albert-László (2018). Science of science. *Science*, **359**(6379).

[147] Fox-Brewster, Tom (September 29, 2014). Londoners give up eldest children in public wi-fi security horror show. https://www.theguardian.com/technology/2014/sep/29/londoners-wi-fi-security-herod-clause. The Guardian, Online, accessed 1 May 2020.

[148] Francis, Enjoli (24 February 2016). Exclusive: Apple CEO Tim Cook says iPhone-cracking software 'equivalent of cancer'. https://abcnews.go.com/Technology/exclusive-apple-ceo-tim-cook-iphone-cracking-software/story?id=37173343. ABC News, Online, accessed 7 October 2019.

[149] Francis, Tracy and Hoefel, Fernanda (2018, November). 'true gen': Generation Z and its implications for companies. Technical report, McKinsey & Company.

[150] Fraser, Colin (4 January 2020). Target didn't figure out a teenager was pregnant before her father did, and that one article that said they did was silly and bad. https://medium.com/@colin.fraser/target-didnt-figure-out-a-teen-girl-was-pregnant-before-her-father-did-a6be13b973a5. Medium, Online, accessed 1 June 2020.

[151] Frey, Carl Benedikt and Osborne, Michael A. (2017). The future of employment: How susceptible are jobs to computerisation? *Technological Forecasting and Social Change*, **114**, 254–280.

[152] Friedler, Sorelle A., Scheidegger, Carlos, Venkatasubramanian, Suresh, Choudhary, Sonam, Hamilton, Evan P., and Roth, Derek (2019). A comparative study of fairness-enhancing interventions in machine learning. In *Proceedings of the Conference on Fairness, Accountability, and Transparency*, FAT* '19, New York, NY, USA, pp. 329–338. Association for Computing Machinery.

[153] Friedman, Jerome H. (1997, January). On bias, variance, 0/1 loss, and the curse-of-dimensionality. *Data Mining & Knowledge Discovery*, **1**(1), 55–77.

[154] Friedman, Jerome H. (2001). Greedy function approximation: A gradient boosting machine. *The Annals of Statistics*.

[155] Fung, Benjamin C. M., Wang, Ke, Chen, Rui, and Yu, Philip S. (2010, June). Privacy-preserving data publishing: A survey of recent developments. *ACM Comput. Surv.*, **42**(4).

[156] Gail Kent (7 May 2018). Hard questions: Why does Facebook enable end-to-end encryption? https://about.fb.com/news/2018/05/end-to-end-encryption/. Facebook Newsroom, Online, accessed 5 October 2019.

[157] Gallagher, Ryan (8 August 2018). Inside Google's effort to develop a censored search engine in China. https://theintercept.com/2018/08/08/google-censorship-china-blacklist/. The Intercept, Online, accessed 1 October 2020.

[158] Gallagher, Ryan (17 December 2018). Google's secret China project "effectively ended" after internal confrontation. https://theintercept.com/2018/12/17/google-china-censored-search-engine-2/. The Intercept, Online, accessed 1 October 2020.

[159] Gallagher, Ryan (9 October 2018). Leaked transcript of private meeting contradicts Google's official story on China. https://theintercept.com/2018/08/08/google-censorship-china-blacklist/. The Intercept, Online, accessed 1 October 2020.

[160] Galloway, Scott (2018). *The Four*. Corgi Books.

[161] Gayo-Avello, Daniel (2012, 04). "I wanted to predict elections with Twitter and all I got was this lousy paper" – a balanced survey on election prediction using Twitter data. arXiv:1204.6441 [cs.CY].

[162] Gelman, Andrew and Loken, Eric (2014, 11). The statistical crisis in science. *American Scientist*, **102**, 460.

[163] Gentry, Craig and Boneh, Dan (2009). *A Fully Homomorphic Encryption Scheme*. Ph.D. thesis, Stanford University.

[164] Gladstone, Joe J., Matz, Sandra C., and Lemaire, Alain (2019). Can psychological traits be inferred from spending? evidence from transaction data. *Psychological Science*, **30**(7), 1087–1096.

[165] Godfrey-Smith, Peter (2003). *Theory and Reality: An Introduction to the Philosophy of Science*. University of Chicago Press Chicago.

[166] Goldman, Megan (2008). Statistics for bioinformatics. https://www.stat.berkeley.edu/~mgoldman/Section0402.pdf. Spring 2008 - Stat C141/ Bioeng C141 - Statistics for Bioinformatics, Online, accessed 8 March 2021.

[167] Goldreich, Oded, Micali, Silvio, and Wigderson, Avi (1991, July). Proofs that yield nothing but their validity or all languages in NP have zero-knowledge proof systems. *J. ACM*, **38**(3), 690–728.

[168] Goldstein, Alex, Kapelner, Adam, Bleich, Justin, and Pitkin, Emil (2015). Peeking inside the black box: Visualizing statistical learning with plots of individual conditional expectation. *Journal of Computational and Graphical Statistics*, **24**(1), 44–65.

[169] Goldwasser, S, Micali, S, and Rackoff, C (1985). The knowledge complexity of interactive proof-systems. In *Proceedings of the Seventeenth Annual ACM Symposium on Theory of Computing*, STOC '85, New York, NY, USA, pp. 291–304. Association for Computing Machinery.

[170] Goldwasser, Shafi, Micali, Silvio, and Rackoff, Charles (1989). The knowledge complexity of interactive proof systems. *SIAM Journal on Computing*, **18**(1), 186–208.

[171] Google. Artificial intelligence at Google: Our principles. https://ai.google/principles. Google, Online, accessed 1 May 2020.

[172] Google (2020). Google mission statement. https://about.google/. Google, Online, accessed 1 October 2020.

[173] Google (2020). Machine learning glossary. https://developers.google.com/machine-learning/glossary/fairness. Google Machine Learning, Online, accessed 1 October 2020.

[174] Google Arts & Culture. Art selfie. https://artsandculture.google.com/camera/selfie. Online, accessed 1 June 2020.

[175] Google Security Blog (30 October 2014). Learning statistics with privacy, aided by the flip of a coin. https://security.googleblog.com/2014/10/learning-statistics-with-privacy-aided.html. Google Security Blog, Online, accessed 1 October 2020.

[176] Graham-Harrison, Emma and Cadwalladr, Carole (17 March 2018). Revealed: 50 million Facebook profiles harvested for Cambridge Analytica in major data breach. https://www.theguardian.com/news/2018/mar/17/cambridge-analytica-facebook-influence-us-election. The Guardian, Online, accessed 7 March 2019.

[177] Greenberg, Andy (13 June 2016). Apple's 'differential privacy' is about collecting your data but not your data. https://www.wired.com/2016/06/apples-differential-privacy-collecting-data/. Wired, Online, accessed 1 October 2020.

[178] Greene, Travis, Martens, David and Shmueli, Galit (2021). Barriers for Academic Data Science Research in the New Realm of Behavior Modification by Digital Platforms. Available at SSRN: https://ssrn.com/abstract=3946116

[179] Greengard, Samuel (2019, December). Will deepfakes do deep damage? *Commuications of the ACM*, **63**(1), 17–19.

[180] Greenspan, Jesse (7 May 2015). The rise and fall of smallpox. https://www.history.com/news/the-rise-and-fall-of-smallpox. History, Online, accessed 1 May 2020.

[181] Grierson, Jamie (11 August 2017). Ex-MI5 chief warns against crackdown on encrypted messaging apps. https://www.theguardian.com/technology/2017/aug/11/ex-mi5-chief-warns-against-crackdown-encrypted-messaging-apps. The Guardian, Online, accessed 1 May 2020.

[182] Grossman, Lev (15 December 2020). Person of the year 2010: Mark zuckerberg. http://content.time.com/time/specials/packages/article/0,28804,2036683_2037183_2037185,00.html. Online, accessed 25 May 2021.

[183] Guardian staff (14 September 2014). Russia's eavesdropping on phone calls examined by Strasbourg Court. https://www.theguardian.com/world/2014/sep/24/strasbourg-court-human-rights-russia-eavesdropping-texts-emails-fsb-. The Guardian, Online, accessed 4 August 2019.

[184] Guidotti, Riccardo, Monreale, Anna, Ruggieri, Salvatore, Turini, Franco, Giannotti, Fosca, and Pedreschi, Dino (2018, August). A survey of methods for explaining black box models. *ACM Comput. Surv.*, **51**(5), 93:1–93:42.

[185] Gunsalus, C.K. and Robinson, Aaron D. (2018). Nine pitfalls of research misconduct. *Nature*, **557**, 297–299.

[186] Hafner, Katie (23 August 2006). Researchers yearn to use aol logs, but they hesitate. https://www.nytimes.com/2006/08/23/technology/23search.html. New York Times, Online, accessed 1 May 2020.

[187] Hafner, Katie and Richtel, Matt (20 January 2006). Google resists U.S. subpoena of search data. https://www.nytimes.com/2006/01/20/technology/google-resists-us-subpoena-of-search-data.html. New York Times, Online, accessed 1 May 2020.

[188] Hansell, Saul (8 August 2006). AOL removes search data on group of web users. https://www.nytimes.com/2006/08/08/business/media/08aol.html. New York Times, Online, accessed 1 May 2020.

[189] Hardt, Moritz, Price, Eric, Price, Eric, and Srebro, Nati (2016). Equality of opportunity in supervised learning. In *Advances in Neural Information Processing Systems 29* (ed. D. D. Lee, M. Sugiyama, U. V. Luxburg, I. Guyon, and R. Garnett), pp. 3315–3323. Curran Associates, Inc.

[190] Harkness, Jon, Lederer, Susan, and Wikler, Daniel (2001, 02). Laying ethical foundations for clinical research. *Bulletin of the World Health Organization*, **79**(4), 365–366.

[191] Harvard Law Today (27 June 2019). Harvard law professor plays instrumental role in creation of facebook's content oversight board. https://today.law.harvard.edu/harvard-law-professor-plays-instrumental-role-in-creation-of-facebooks-content-oversight-board/. Online, accessed 25 May 2021.

[192] Hashemian, H. M. and Bean, Wendell C. (2011). State-of-the-art predictive maintenance techniques. *IEEE Transactions on Instrumentation and Measurement*, **60**(10), 3480–3492.

[193] Hawes, Michael (5 March 2020). Differential privacy and the 2020 decennial census. https://www2.census.gov/about/policies/2020-03-05-differential-privacy.pdf. U.S. Census Bureau, Online, accessed 1 October 2020.

[194] Hawkins, Andrew J. (9 December 2019). Waymo's driverless car: Ghost-riding in the back seat of a robot taxi. https://www.theverge.com/2019/12/9/21000085/waymo-fully-driverless-car-self-driving-ride-hail-service-phoenix-arizona. The Verge, Online, accessed 8 March 2021.

[195] Head, Morgan L., Holman, Luke, Lanfear, Rob, Kahn, Andrew T., and Jennions, Michael D. (2015). The extent and consequences of p-hacking in science. *PLoS Biology*, **13**(3).

[196] Heath, Alex (25 May 2017). Mark Zuckerberg's big Harvard speech was his most political moment yet. https://www.businessinsider.nl/watch-video-transcript-mark-zuckerberg-harvard-commencement-speech-2017-5. Online, accessed 8 March 2021.

[197] Hee, Lente Van, Baert, Denny, Verheyden, Tim, and Heuvel, Ruben Van Den (10 July 2019). Google employees are eavesdropping, even in your living room, VRT NWS has discovered. https://www.vrt.be/vrtnws/en/2019/07/10/google-employees-are-eavesdropping-even-in-flemish-living-rooms/. VRT NWS, Online, accessed 28 February 2020.

[198] Helen Nissenbaum (1998). Protecting privacy in an information age: The problem of privacy in public. *Law and Philosophy*, **17**, 559–596.

[199] Hern, Alex (10 April 2018). How to check whether Facebook shared your data with Cambridge Analytica. https://www.theguardian.com/technology/2018/apr/10/facebook-notify-users-data-harvested-cambridge-analytica. The Guardian, Online, accessed 7 March 2019.

[200] Hern, Alex (27 June 2014). New York taxi details can be extracted from anonymised data, researchers say. https://www.theguardian.com/technology/2014/jun/27/new-york-taxi-details-anonymised-data-researchers-warn. The Guardian, Online, accessed 1 May 2020.

[201] Hicks, Michael and Devaraj, Srikant (2012, June). the myth and the reality of of manufacturing in America. Technical report, Ball State University.

[202] High-Level Expert Group on Artificial Intelligence (2019). Ethics guidelines for trustworthy AI. Technical report, European Commission.

[203] High-Level Expert Group on Artificial Intelligence (2019). Ethics guidelines for trustworty AI. Technical report, European Commission.

[204] Hill, Austin Bradford (1963). Medical ethics and controlled trials. *British medical journal*, **1**(5337), 1043–1049.

[205] Hill, Haskmir (28 July 2014). Okcupid lied to users about their compatibility as an experiment. https://www.forbes.com/sites/kashmirhill/2014/07/28/okcupid-experiment-compatibility-deception/?sh=5b653ffa77b1. Online, accessed 25 May 2021.

[206] Hill, Kashmir (16 February 2012). How target figured out a teen girl was pregnant before her father did. https://www.forbes.com/sites/kashmirhill/2012/02/16/how-target-figured-out-a-teen-girl-was-pregnant-before-her-father-did/#454c5d536668. Forbes, Online, accessed 1 June 2020.

[207] Hill, Kashmir (18 January 2020). The secretive company that might end privacy as we know it. https://www.nytimes.com/2020/01/18/technology/clearview-privacy-facial-recognition.html. New York Times, Online, October June 1, 2020.

[208] Hill, Kashmir (28 June 2014). Facebook manipulated 689,003 users' emotions for science. https://www.forbes.com/sites/kashmirhill/2014/06/28/facebook-manipulated-689003-users-emotions-for-science/. Online, accessed 25 May 2021.

[209] Hill, Kashmir (5 March 2020). Before clearview became a police tool, it was a secret plaything of the rich. https://www.nytimes.com/2020/03/05/technology/clearview-investors.html. New York Times, Online, accessed 1 October 2020.

[210] Hillier, Amy E. (2003). Redlining and the Homeowners' Loan Corporation. Departmental Papers (City and Regional Planning) 5-1-2003, University of Pennsylvania. Postprint version. Published in *Journal of Urban History* 29(4), 394–420, 2003.

[211] Hopwood, D., Bowe, S., Hornby, T., and Wilcox, N. (2020). Zcash protocol specification. White Paper Version 2020.1.14, Electric Coin Company.

[212] Horizon 2020 Commission Expert Group to advise on specific ethical issues raised by driverless mobility (E03659) (2020). Ethics of connected and automated vehicles: recommendations on road safety, privacy, fairness, explainability and responsibility. https://ec.europa.eu/info/sites/info/files/research_and_innovation/ethics_of_connected_and_automated_vehicles_report.pdf. Publication Office of the European Union: Luxembourg, Online, accessed 21 January 2021.

[213] Hsu, Justin, Gaboardi, Marco, Haeberlen, Andreas, Khanna, Sanjeev, Narayan, Arjun, Pierce, Benjamin C., and Roth, Aaron (2014). Differential privacy: An economic method for choosing epsilon. In *Proceedings of the 2014 IEEE 27th Computer Security Foundations Symposium*, CSF '14, USA, pp. 398–410. IEEE Computer Society.

[214] Hsu, Tiffany (22 April 2020). An ESPN commercial hints at advertising's deepfake future. https://www.nytimes.com/2020/04/22/business/media/espn-kenny-mayne-state-farm-commercial.html. Online, accessed 25 May 2021.

[215] Human Rights Library (1992). All amendments to the united states constitution. http://hrlibrary.umn.edu/education/all_amendments_usconst.htm. Online, accessed 25 May 2021.

[216] Hunt, Elle (24 March 2016). Tay, Microsoft's AI chatbot, gets a crash course in racism from Twitter. https://www.theguardian.com/technology/2016/mar/24/tay-microsofts-ai-chatbot-gets-a-crash-course-in-racism-from-twitter. The Guardina, Online, accessed 1 October 2020.

[217] Huysmans, Johan, Dejaeger, Karel, Mues, Christophe, Vanthienen, Jan, and Baesens, Bart (2011, 04). An empirical evaluation of the comprehensibility of decision table, tree and rule based predictive models. *Decision Support Systems*, **51**, 141–154.

[218] Hvistendahl, Mara (2013). China's publication bazaar. *Science*, **342**(6162), 1035–1039.

[219] Ingram, Davud (5 May 2021). Trump's facebook ban upheld by oversight board. https://www.nbcnews.com/tech/tech-news/trump-s-facebook-ban-upheld-oversight-board-n1266339. Online, accessed 25 May 2021.

[220] Iosifidis, Vasileios and Ntoutsi, Eirini (2019). Adafair: Cumulative fairness adaptive boosting. In *Proceedings of the 28th ACM International Conference on Information and Knowledge Management*, CIKM '19, New York, NY, USA, pp. 781–790. Association for Computing Machinery.

[221] Isaac, Mike (23 April 2017). Uber's C.E.O. plays with fire. https://www.nytimes.com/2017/04/23/technology/travis-kalanick-pushes-uber-and-himself-to-the-precipice.html. New York Times, Online, accessed 1 October 2020.

[222] Iskyan, Kim (2 May 2016). Sell in May and don't come back until November. https://www.businessinsider.com/sell-in-may-dont-come-back-until-november-2016-5. Online, accessed 8 March 2021.

[223] Jackson, Dan (7 July 2017). The Netflix prize: How a $1 million contest changed binge-watching forever. https://www.thrillist.com/entertainment/nation/the-netflix-prize. Thrillist, Online, accessed 1 May 2020.

[224] Jain, Atish, Dedhia, Ronak, and Patil, Abhijit (2015, Nov). Enhancing the security of caesar cipher substitution method using a randomized approach for more secure communication. *International Journal of Computer Applications*, **129**(13), 6–11.

[225] James B. Comey (16 October 2014). Going dark: Are technology, privacy, and public safety on a collision course? https://www.fbi.gov/news/speeches/going-dark-are-technology-privacy-and-public-safety-on-a-collision-course. FBI, Online, accessed 13 September 2019.

[226] Junque de Fortuny, Enric, De Smedt, Tom, Martens, David, and Daelemans, Walter (2014, 03). Evaluating and understanding text-based stock price prediction models. *Information Processing & Management*, **50**, 426–441.

[227] Junqué de Fortuny, Enric, Martens, David, and Provost, Foster (2013). Predictive modeling with big data: is bigger really better? *Big Data*, **1**(4), 215–226.

[228] Junqué de Fortuny, Enric, Stankova, Marija, Moeyersoms, Julie, Minnaert, Bart, Provost, Foster, and Martens, David (2014). Corporate residence fraud detection. In *Proceedings of the 20th ACM SIGKDD international conference on Knowledge discovery and data mining*, pp. 1650–1659. ACM.

[229] Jurafsky, Daniel and Martin, James H. (2000). *Speech and Language Processing: An Introduction to Natural Language Processing, Computational Linguistics and Speech Recognition (Prentice Hall Series in Artificial Intelligence)* (1 edn). Prentice Hall.

[230] Kairouz, Peter, McMahan, H. Brendan, Avent, Brendan, Bellet, Aurélien, Bennis, Mehdi, Bhagoji, Arjun Nitin, Bonawitz, Kallista, Charles, Zachary, Cormode, Graham, Cummings, Rachel, D'Oliveira, Rafael G. L., Eichner, Hubert, Rouayheb, Salim El, Evans, David, Gardner, Josh, Garrett, Zachary, Gascón, Adrià, Ghazi, Badih, Gibbons, Phillip B., Gruteser, Marco, Harchaoui, Zaid, He, Chaoyang, He, Lie, Huo, Zhouyuan, Hutchinson, Ben, Hsu, Justin, Jaggi, Martin, Javidi, Tara, Joshi, Gauri, Khodak, Mikhail, Konečný, Jakub, Korolova, Aleksandra, Koushanfar, Farinaz, Koyejo, Sanmi, Lepoint, Tancrède, Liu, Yang, Mittal, Prateek, Mohri, Mehryar, Nock, Richard, Özgür, Ayfer, Pagh, Rasmus, Raykova, Mariana, Qi, Hang, Ramage, Daniel, Raskar, Ramesh, Song, Dawn, Song, Weikang, Stich, Sebastian U., Sun, Ziteng, Suresh, Ananda Theertha, Tramèr, Florian, Vepakomma, Praneeth,

Wang, Jianyu, Xiong, Li, Xu, Zheng, Yang, Qiang, Yu, Felix X., Yu, Han, and Zhao, Sen (2021). Advances and open problems in federated learning. arXiv:1912.04977 [cs.LG].

[231] Kalai, Gil (2016). The quantum computer puzzle. *Notices of the AMS*, **63**(5), 508–516.

[232] Kamber, Scott A., Malley, Joseph H., and Parisi, David (17 December 2009). Law suit of doe v. neflix. https://www.wired.com/images_blogs/threatlevel/2009/12/doe-v-netflix.pdf. Online, accessed 25 May 2021.

[233] Kamiran, Faisal and Calders, Toon (2012, October). Data preprocessing techniques for classification without discrimination. *Knowl. Inf. Syst.*, **33**(1), 1–33.

[234] Kamiran, Faisal, Calders, Toon, and Pechenizkiy, Mykola (2010). Discrimination aware decision tree learning. In *Proceedings of the 2010 IEEE International Conference on Data Mining*, ICDM '10, USA, pp. 869–874. IEEE Computer Society.

[235] Kamishima, Toshihiro, Akaho, Shotaro, Asoh, Hideki, and Sakuma, Jun (2012). Fairness-aware classifier with prejudice remover regularizer. In *Machine Learning and Knowledge Discovery in Databases – European Conference, ECML PKDD 2012, Bristol, UK, September 24-28, 2012. Proceedings, Part II* (ed. P. A. Flach, T. D. Bie, and N. Cristianini), Volume 7524 of *Lecture Notes in Computer Science*, pp. 35–50. Springer.

[236] Kang, Cecilia (12 July 2019). F.T.C. approves facebook fine of about $5 billion. https://www.nytimes.com/2019/07/12/technology/facebook-ftc-fine.html. Online, accessed 25 May 2021.

[237] Karimi, Amir-Hossein, Barthe, Gilles, Schölkopf, Bernhard, and Valera, Isabel (2021). A survey of algorithmic recourse: definitions, formulations, solutions, and prospects. arXiv:2010.04050 [cs.LG].

[238] Kasperkevic, Jana (1 July 2015). Google says sorry for racist auto-tag in photo app. https://www.theguardian.com/technology/2015/jul/01/google-sorry-racist-auto-tag-photo-app. Online, accessed 25 May 2021.

[239] Kayande, U., De Bruyn, A., Lilien, G. L., Rangaswamy, A., and van Bruggen, G. H. (2009). How incorporating feedback mechanisms in a DSS affects dss evaluations. *Information Systems Research*, **20**, 527–546.

[240] Keynes, John Maynard (1930). Economic possibilities for our grandchildren. In *Essays in Persuasion*, pp. 358–373. Harcourt Brace.

[241] Kleinberg, Jon, Ludwig, Jens, Mullainathan, Sendhil, and Rambachan, Ashesh (2018, May). Algorithmic fairness. *AEA Papers and Proceedings*, **108**, 22–27.

[242] Kleinberg, Jon M., Mullainathan, Sendhil, and Raghavan, Manish (2017). Inherent trade-offs in the fair determination of risk scores. In *8th Innovations in Theoretical Computer Science Conference, ITCS 2017, January 9-11, 2017, Berkeley, CA, USA* (ed. C. H. Papadimitriou), Volume 67 of *LIPIcs*, pp. 43:1–43:23. Schloss Dagstuhl – Leibniz-Zentrum für Informatik.

[243] Kodratoff, Yves (1994). The comprehensibility manifesto. *KDD Nuggets*, **94**(9).

[244] Kofman, Ava and Tobin, Ariana (13 December 2019). Facebook ads can still discriminate against women and older workers, despite a civil rights settlement. https://www.propublica.org/article/facebook-ads-can-still-discriminate-against-women-and-older-workers-despite-a-civil-rights-settlement. Online, accessed 25 May 2021.

[245] Konečný, Jakub, McMahan, H. Brendan, Yu, Felix X., Richtárik, Peter, Suresh, Ananda Theertha, and Bacon, Dave (2016). Federated learning: Strategies for improving communication efficiency. In *NIPS Workshop on Private Multi-Party Machine Learning*.

[246] Kosinski, Michal, Stillwell, David, and Graepel, Thore (2013). Private traits and attributes are predictable from digital records of human behavior. *Proceedings of the National Academy of Sciences*, **110**(15), 5802–5805.

[247] Kramer, Adam D. I., Guillory, Jamie E., and Hancock, Jeffrey T. (2014). Experimental evidence of massive-scale emotional contagion through social networks. *Proceedings of the National Academy of Sciences*, **111**(24), 8788–8790.

[248] Kranzberg, Melvin (1995). Technology and history: 'Kranzberg's laws'. *Bulletin of Science, Technology & Society*, **15**(1), 5–13.

[249] Kravets, David (17 March 2010). Judge approves $9.5 million Facebook 'Beacon' accord. https://www.wired.com/2010/03/facebook-beacon-2/. Wired, Online, accessed 1 June 2020.

[250] Kremp, Matthias (22 November 2018). Knuddle's chat platform must pay a fine after hacking. https://www.spiegel.de/netzwelt/web/knuddels-chat-plattform-muss-nach-hackerangriff-bussgeld-zahlen-a-1239776.html. Spiegel, Online, accessed 19 July 2019.

[251] Kroll, Joshua A. (2018). Data science data governance [AI ethics]. *IEEE Security Privacy*, **16**(6), 61–70.

[252] Krysia Lenzo and Anita Balakrishnan (29 February 2016). Chertoff: Apple's right, the internet has changed. https://www.cnbc.com/2016/02/29/apple-attorney-fbis-request-is-concerning.html. CNBC, Online, accessed 13 September 2019.

[253] Kshetri, Nir (2014). Big data's impact on privacy, security and consumer welfare. *Telecommunications Policy*, **38**(11), 1134–1145.

[254] Kummer, Wiebke (24 October 2018). Portuguese data protection authority imposes 400,000 € fine on hospital. https://www.datenschutz-notizen.de/portuguese-data-protection-authority-imposes-400000-e-fine-on-hospital-4821441/. Datenschutz Notizen, Online, accessed 7 March 2019.

[255] Kursuncu, Ugur, Gaur, Manas, Lokala, Usha, Thirunarayan, Krishnaprasad, Sheth, Amit, and Arpinar, I. Budak (2018). Predictive analysis on twitter: Techniques and applications. arXiv:1806.02377 [cs.SI].

[256] Kusner, Matt J, Loftus, Joshua, Russell, Chris, and Silva, Ricardo (2017). Counterfactual fairness. In *Advances in Neural Information Processing Systems* (ed. I. Guyon, U. V. Luxburg, S. Bengio, H. Wallach, R. Fergus, S. Vishwanathan, and R. Garnett), Volume 30, pp. 4066–4076. Curran Associates, Inc.

[257] Lacoste, Alexandre, Luccioni, Alexandra, Schmidt, Victor, and Dandres, Thomas (2019). Quantifying the carbon emissions of machine learning. arXiv:1910.09700 [cs.CY].

[258] Lagnado, L.M. and Dekel, S.C. (1992). *Children of the Flames: Dr. Josef Mengele and the Untold Story of the Twins of Auschwitz*. Penguin Books.

[259] Lawrence, Chris (11 February 2020). Vestager (EU): "we must ensure transparency with the data that is processed and its use". https://sportsfinding.com/vestager-eu-we-must-ensure-transparency-with-the-data-that-is-processed-and-its-use/17488/. Sportsfinding, Online, accessed 1 June 2020.

[260] Lazzaro, Sage (17 August 2017). Is this soap dispenser racist? controversy as Facebook employee shares video of machine that only responds to white skin. https://www.dailymail.co.uk/sciencetech/article-4800234/Is-soap-dispenser-RACIST.html. Daily Mail, Online, accessed 28 February 2020.

[261] Le Monde (4 November 1999). La vie privée du chef de l'etat et alors ? https://www.lemonde.fr/archives/article/1994/11/04/la-vie-privee-du-chef-de-l-etat-et-alors_3849431_1819218.html. Online, accessed 25 May 2021.

[262] LeCun, Yann, Bengio, Yoshua, and Hinton, Geoffrey (2015). Deep learning. *Nature*, **521**, 436–444.

[263] Lee, Peter (25 March 2016). Learning from tay's introduction. https://blogs.microsoft.com/blog/2016/03/25/learning-tays-introduction/. Official Microsoft Blog, Online, accessed 1 October 2020.

[264] Legal Information Institute. 15 U.S. code § 1691 - scope of prohibition. https://www.law.cornell.edu/uscode/text/15/1691. Cornell Law School, Online, accessed 28 October 2019.

[265] Leonard, John, Mindelle, David, and Stayton, Erik (2020). Autonomous vehicles, mobility, and employment policy: The roads ahead.

[266] Levin, Sam (26 July 2018). Amazon face recognition falsely matches 28 lawmakers with mugshots, ACLU says. https://www.theguardian.com/technology/2018/jul/26/amazon-facial-rekognition-congress-mugshots-aclu. theguardian.com, Online, accessed 1 June 2020.

[267] LfDI Baden-Württemberg (November 22, 2018). LfDI Baden-Württemberg imposes its first fine in Germany according to the GDPR. https://www.baden-wuerttemberg.datenschutz.de/lfdi-baden-wuerttemberg-verhaengt-sein-erstes-bussgeld-in-deutschland-nach-der-ds-gvo/. Online, translated from German, accessed 25 May 2021.

[268] Li, N., Li, T., and Venkatasubramanian, S. (2007). t-closeness: Privacy beyond k-anonymity and l-diversity. In *2007 IEEE 23rd International Conference on Data Engineering*, pp. 106–115.

[269] Liptak, Andrew (23 April 2017). Uber tried to fool Apple and got caught. https://www.theverge.com/2017/4/23/15399438/apple-uber-app-store-fingerprint-program-tim-cook-travis-kalanick. The Verge, Online, accessed 1 October 2020.

[270] Logg, Jennifer M., Minson, Julia A., , and Moore, Don A. (26 October 2018). Do people trust algorithms more than companies realize? https://hbr.org/2018/10/do-people-trust-algorithms-more-than-companies-realize. Online, accessed 25 May 2021.

[271] Lohr, Steve (12 March 2010). Netflix cancels contest after concerns are raised about privacy. https://www.nytimes.com/2010/03/13/technology/13netflix.html. New York Times, Online, accessed 1 May 2020.

[272] Los Angeles Times (9 December 2015). San Bernardino shooting updates. https://www.latimes.com/local/lanow/la-me-ln-san-bernardino-shooting-live-updates-htmlstory.html. Online, accessed 25 May 2021.

[273] Lundberg, Scott M. and Lee, Su-In (2017). A unified approach to interpreting model predictions. In *Proceedings of the 31st International Conference on Neural Information Processing Systems*, NIPS'17, Red Hook, NY, USA, pp. 4768–4777. Curran Associates Inc.

[274] Machanavajjhala, Ashwin, Kifer, Daniel, Gehrke, Johannes, and Venkitasubramaniam, Muthuramakrishnan (2007, March). l-diversity: Privacy beyond k-anonymity. *ACM Trans. Knowl. Discov. Data*, **1**(1), 3–es.

[275] Madiega, Tambiama (2019). *EU Guidelines on Ethics in Artificial Intelligence: Context and Implementation*. Technical report, European Parliamentary Research Service.

[276] Madry, Aleksander, Makelov, Aleksandar, Schmidt, Ludwig, Tsipras, Dimitris, and Vladu, Adrian (2018). Towards deep learning models resistant to adversarial attacks. In *6th International Conference on Learning Representations, ICLR 2018, Vancouver, BC, Canada, April 30–May 3, 2018, Conference Track Proceedings*. OpenReview.net.

[277] Maimon, Oded and Rokach, Lior (2005, 01). Decomposition methodology for knowledge discovery and data mining. theory and applications. In: Maimon O., Rokach L. (eds) Data Mining and Knowledge Discovery Handbook. Springer, Boston, MA.

[278] Manthorpe, Rowland and Martin, Alexander J. (4 July 2019). 81% of 'suspects' flagged by MET's police facial recognition technology innocent, independent report says. https://news.sky.com/story/met-polices-facial-recognition-tech-has-81-error-rate-independent-report-says-11755941. news.sky.com, Online, accessed 1 June 2020.

[279] Manyika, James, Chui, Michael, Brown, Brad, Bughin, Jacques, Dobbs, Richard, Roxburgh, Charles, and Byers, Angela Hung (2011, June). Big data: The next frontier for innovation, competition, and productivity. Technical report, McKinsey Global Institute.

[280] Manyika, James, Lund, Susan, Chui, Michael, Bughin, Jacques, Woetzel, Jonathan, Batra, Parul, Ko, Ryan, and Sanghvi, Saurabh (2017). Jobs lost, jobs gained: Workforce transitions in a time of automation. Online, accessed 8 March 2021.

[281] Map Happy (16 December 2014). Use a 'fake' location to get cheaper plane tickets. https://www.huffpost.com/entry/use-a-fake-location-to-ge_b_6315424. Online, accessed 8 March 2021.

[282] Maréchal, Nathalie (2017). Networked authoritarianism and the geopolitics of information: Understanding Russian internet policy. *Media and Communication*, **5**(1), 29–41.

[283] Marshall, Aarian (8 December 2018). 32 hours in Chandler, Arizona, the self-driving capital of the world. https://www.wired.com/story/32-hours-chandler-arizona-self-driving-capital/. wired.com, Online, accessed 8 March 2021.

[284] Martens, David (2008, December). Building acceptable classification models for financial engineering applications: Thesis summary. *SIGKDD Explor. Newsl.*, **10**(2), 30–31.

[285] Martens, David, Baesens, Bart, and Van Gestel, Tony (2009). Decompositional rule extraction from support vector machines by active learning. *IEEE Transactions on Knowledge and Data Engineering*, **21**(2), 178–191.

[286] Martens, David, Baesens, Bart, Van Gestel, Tony, and Vanthienen, Jan (2007). Comprehensible credit scoring models using rule extraction from support vector machines. *European Journal of Operational Research*, **183**(3), 1466–1476.

[287] Martens, David and Provost, Foster (2014, March). Explaining data-driven document classifications. *MIS Quarterly*, **38**(1), 73–100.

[288] Martens, David, Vanthienen, Jan, Verbeke, Wouter, and Baesens, Bart (2011). Performance of classification models from a user perspective. *Decision Support Systems*, **51**(4), 782–793.

[289] Martin, Alicia R., Kanai, Masahiro, Kamatani, Yoichiro, Okada, Yukinori, Neale, Benjamin M., and Daly, Mark J. (2019, 4). Clinical use of current polygenic risk scores may exacerbate health disparities. *Nature Genetics*, **51**(4), 584–591.

[290] Martinez, Antonio Garcia (6 November 2019). Are Facebook ads discriminatory? it's complicated. https://www.wired.com/story/are-facebook-ads-discriminatory-its-complicated/. Online, accessed 25 May 2021.

[291] Marwick, Alice and Lewis, Rebecca (2017). *Media Manipulation and Disinformation Online*.

[292] Mattioli, Dana (23 August 2012). On Orbitz, Mac users steered to pricier hotels. https://www.wsj.com/articles/SB10001424052702304458604577488822667325882. Online, accessed 8 March 2021.

[293] Matz, S. C., Kosinski, M., Nave, G., and Stillwell, D. J. (2017). Psychological targeting as an effective approach to digital mass persuasion. *Proceedings of the National Academy of Sciences*, **114**(48), 12714–12719.

[294] McCaskill, Steve (17 March 2018). La liga handed $ 280,000 GDPR fine for 'spying' on fans watching pirated streams. https://www.forbes.com/sites/stevemccaskill/2019/06/12/la-liga-handed-e250000-gdpr-fine-for-spying-on-fans-watching-pirated-streams. Forbes, Online, accessed 7 March 2019.

[295] McMahan, Brendan, Moore, Eider, Ramage, Daniel, Hampson, Seth, and y Arcas, Blaise Aguera (2017, 20–22 April). Communication-Efficient Learning of Deep Networks from Decentralized Data. Volume 54 of *Proceedings of Machine Learning Research*, Fort Lauderdale, FL, USA, pp. 1273–1282. PMLR.

[296] Merali, Zeeya (2015). Quantum 'spookiness' passes toughest test yet. *Nature*, **525**, 14–15.

[297] Merton, Robert K. (1957). Priorities in scientific discovery: A chapter in the sociology of science. *American Sociological Review*, **22**(6), 635–659.

[298] Meyerson, Adam and Williams, Ryan (2004). On the complexity of optimal k-anonymity. In *Proceedings of the Twenty-Third ACM SIGMOD-SIGACT-SIGART Symposium on Principles of Database Systems*, PODS '04, New York, NY, USA, pp. 223–228. Association for Computing Machinery.

[299] Michalski, Ryszard (1983). A theory and methodology of inductive learning. *Artificial Intelligence*, **20**(2), 111–161.

[300] Mikhitarian, Sarah (25 April 2018). Home values remain low in vast majority of formerly redlined neighborhoods. https://www.zillow.com/research/home-values-redlined-areas-19674/. Zillow Research, Online, accessed 21 January 2021.

[301] Miller, A. Ray (1995). The cryptographic mathematics of enigma. *Cryptologia*, **19**(1), 65–80.

[302] Miller, G.A. (1956). The magical number seven, plus or minus two: some limits to our capacity for processing information. *Psychological review*, **63**(2), 81–97.

[303] Milojevic, Stasa (2015). Quantifying the cognitive extent of science. *Journal of Informetrics*, **9**(4), 962–973.

[304] Mislove, Alan, Lehmann, Sune, Ahn, Yong-Yeol, Onnela, Jukka-Pekka, and Rosenquist, (James) (2011), (01). Understanding the demographics of twitter users. Volume 11.

[305] Mitchell, Margaret, Wu, Simone, Zaldivar, Andrew, Barnes, Parker, Vasserman, Lucy, Hutchinson, Ben, Spitzer, Elena, Raji, Inioluwa Deborah, and Gebru, Timnit (2019, Jan). Model cards for model reporting. *Proceedings of the Conference on Fairness, Accountability, and Transparency*.

[306] Moeyersoms, Julie, d'Alessandro, Brian, Provost, Foster, and Martens, David (2016). Explaining classification models built on high-dimensional sparse data.

[307] Molnar, Christoph (2019). *Interpretable Machine Learning*. https://christophm.github.io/interpretable-ml-book/.

[308] Mongelli, Lorena (March 16, 2012). NYPD uses high-tech facial-recognition software to nab barbershop shooting suspect. http://nypost.com/2012/03/16/nypd-uses-high-tech-facial-recognition-software-to-nab-barbershop-shooting-suspect/. New York Post, Online, accessed 1 June 2020.

[309] Monteiro, Ana Menezes (3 January 2019). First GDPR fine in portugal issued against hospital for three violations. https://iapp.org/news/a/first-gdpr-fine-in-portugal-issued-against-hospital-for-three-violations/. IAPP, Online, accessed 7 March 2019.

[310] Morais, Eduardo, Koens, Tommy, van Wijk, Cees, and Koren, Aleksei (2019). A survey on zero knowledge range proofs and applications. *CoRR*, abs/1907.06381.

[311] Mulgan, Richard (2000). 'accountability': An ever-expanding concept? *Public Administration*, **78**(3), 555–573.

[312] Murphy, Hannah (17 May 2021). Facebook's oversight board: an imperfect solution to a complex problem. https://www.ft.com/content/802ae18c-af43-437b-ae70-12a87c838571. Online, accessed 25 May 2021.

[313] Naor, Moni, Naor, Yael, and Reingold, Omer (1999). Applied kid cryptography or how to convince your children you are not cheating. In *In Proc. of Eurocrypt '94*, pp. 1–12.

[314] Narayanan, Arvind and Shmatikov, Vitaly (2006). How to break anonymity of the Netflix prize dataset.

[315] Narayanan, Arvind and Shmatikov, Vitaly (2008). Robust de-anonymization of large sparse datasets. In *Proceedings of the 2008 IEEE Symposium on Security and Privacy*, SP '08, USA, pp. 111–125. IEEE Computer Society.

[316] Narayanan, Arvind and Shmatikov, Vitaly (2010, June). Myths and fallacies of 'personally identifiable information'. *Communications of the ACM*, **53**(6), 24–26.

[317] Narla, Akhila, Kuprel, Brett, Sarin, Kavita, Novoa, Roberto, and Ko, Justin (2018). Automated classification of skin lesions: From pixels to practice. *Journal of Investigative Dermatology*, **138**(10), 2108–2110.

[318] Natarajan, Sridhar and Nasiripour, Shahien (9 November 2019). Viral tweet about apple card leads to goldman sachs probe. https://www.bloomberg.com/news/articles/2019-11-09/viral-tweet-about-apple-card-leads-to-probe-into-goldman-sachs. Bloomberg, Online, accessed 21 January 2021.

[319] NBC News (28 July 2014). Okcupid experiments on users, discovers everyone is shallow. https://www.nbcnews.com/tech/internet/okcupid-experiments-users-discovers-everyone-shallow-n166761. Online, accessed 25 May 2021.

[320] Netflix (2009). Netflix prize – leaderboard. https://netflixprize.com/leaderboard.html. Netflix, Online, accessed May 1, 2020.

[321] Netflix (July, 2009). Contest closed. https://web.archive.org/web/20090728003547/http://www.netflixprize.com/closed. Netflix, Online, accessed 1 May 2020.

[322] News, BBC (17 July 2019). Google's project dragonfly 'terminated' in China. https://www.bbc.com/news/technology-49015516. BBC News, Online, accessed 1 October 2020.

[323] News 18 (9 May 2018). Whatsapp's end-to-end encryption is a powerful tool for security and safety: Facebook. https://www.news18.com/news/tech/whatsapps-end-to-end-encryption-is-a-powerful-tool-for-security-and-safety-facebook-1742309.html. News 18, Online, accessed 28 February 2020.

[324] Niederer, Sabine and van Dijck, José (2010). Wisdom of the crowd or technicity of content? Wikipedia as a sociotechnical system. *New Media & Society*, **12**(8), 1368–1387.

[325] Nielsen, Michael A. and Chuang, Isaac L. (2010). *Quantum Computation and Quantum Information*. Cambridge University Press.

[326] Nietzsche, Friedrich (1886). *Beyond Good and Evil*. Vintage.

[327] Nissenbaum, Helen (2011). A contextual approach to privacy online. *Daedalus*, **140**(4), 32–48.

[328] Nissim, Kobbi, Steinke, Thomas, Wood, Alexandra, Altman, Micah, Bembenek, Aaron, Bun, Mark, Gaboardi, Marco, O'Brien, David R., and Vadhan, Salil (2018). Differential privacy: A primer for a non-technical audience. *Privacy Law Scholars Conference 2017*.

[329] NIST (5 August 2015). Nist releases sha-3 cryptographic hash standard. https://www.nist.gov/news-events/news/2015/08/nist-releases-sha-3-cryptographic-hash-standard. National Institute of Standards and Technology (NIST), Online, accessed 19 July 2019.

[330] Nitulescu, Anca (2020). zk-snarks: A gentle introduction. Technical report.

[331] Noll, Michael G. (7 August 2006). AOL research publishes 650,000 user queries. https://www.michael-noll.com/blog/2006/08/07/aol-research-publishes-500k-user-queries/. Blog Michael G. Noll, Online, accessed 1 May 2020.

[332] NPR (2 March 2015). Ben Franklin's famous 'liberty, safety' quote lost its context in 21st century. https://www.npr.org/2015/03/02/390245038/ben-franklins-famous-liberty-safety-quote-lost-its-context-in-21st-century. NPR, Online, accessed 28 February 2020.

[333] O'Connor, Joseph (11 July 2018). Websites not available in the European Union after GDPR. https://data.verifiedjoseph.com/dataset/websites-not-available-eu-gdpr. Online, accessed 1 October 2020.

[334] Official Journal of the European Union (1996). Directive 96/9/ec of the European Parliament and of the council of 11 March 1996 on the legal protection of databases. Technical report, European Parliament and Council.

[335] Official Journal of the European Union (2016). Regulation (EU) 2016/679 of the European Parliament and of the Council of 27 April 2016 on the protection of natural persons with regard to the processing of personal data and on the free movement of such data, and repealing directive 95/46/ec (General Data Protection Regulation). Technical report, European Parliament and Council.

[336] O'Flaherty, Kate (14 December 2018). The worst passwords of 2018 show the need for better practices. https://www.forbes.com/sites/kateoflahertyuk/2018/12/14/these-are-the-top-20-worst-passwords-of-2018. Forbes, Online, accessed 19 July 2019.

[337] Ohlheiser, Abby (25 March 2016). Trolls turned Tay, Microsoft's fun millennial AI bot, into a genocidal maniac. https://www.washingtonpost.com/news/the-intersect/wp/2016/03/24/the-internet-turned-tay-microsofts-fun-millennial-ai-bot-into-a-genocidal-maniac/. Online, accessed 25 May 2021.

[338] Ohm, Paul (2010, 08). Broken promises of privacy: Responding to the surprising failure of anonymization. *UCLA Law Review*, **57**, 1701–1777.

[339] Oversight Board (2021). Introduction – oversight board. https://oversightboard.com/governance/. Online, accessed 25 May 2021.

[340] Oversight Board (2021). Members – oversight board. https://oversightboard.com/governance/. Online, accessed 25 May 2021.

[341] Oversight Border (September, 2019). Oversight board charter. https://about.fb.com/wp-content/uploads/2019/09/oversight_board_charter.pdf. Online, accessed 30 November 2021.

[342] Panzarino, Matthew (14 March 2019). Apple ad focuses on iPhone's most marketable feature — privacy. https://techcrunch.com/2019/03/14/apple-ad-focuses-on-iphones-most-marketable-feature-privacy. Tech Crunch, Online, accessed 28 October 2019.

[343] Pardes, Arielle (24 May 2018). What is GDPR and why should you care? https://www.wired.com/story/how-gdpr-affects-you/. Wired, Online, accessed 28 October 2019.

[344] Penguin Press – Jordan Ellenberg (14 July 2016). Abraham wald and the missing bullet holes. https://medium.com/@penguinpress/an-excerpt-from-how-not-to-be-wrong-by-jordan-ellenberg-664e708cfc3d. medium.com, Online, accessed 7 October 2019.

[345] Perneger, Thomas V (1998). What's wrong with bonferroni adjustments. *BMJ*, **316**(7139), 1236–1238.

[346] Pernot-Leplay, Emmanuel (2020). China's approach on data privacy law: A third way between the U.S. and the EU? *Penn State Journal of Law & International Affairs*, **8**(1).

[347] Pesenti, Jerome (2 November 2021). An Update on our Use of Face Recognition. https://about.fb.com/news/2021/11/update-on-use-of-face-recognition/. Online, accessed 25 November 2021.

[348] Peteranderl, Sonja (1 January 2019). DSGVO fines 'mistakes are now expensive'. https://www.spiegel.de/netzwelt/netzpolitik/dsgvo-strafen-fehler-werden-jetzt-teuer-a-1249443.html. Spiegel, Online, accessed 19 July 2019.

[349] Pew Research Center (12 June 2019). Mobile fact sheet. https://www.pewresearch.org/internet/fact-sheet/mobile/. Pew Research, Online, accessed 11 November 2019.

[350] Pichai, Sundar (7 June 2018). AI at Google: our principles. https://www.blog.google/technology/ai/ai-principles/. Online, accessed 8 March 2021.

[351] Pogge, J. (1992, 01). Reinvestment in Chicago neighborhoods: A twenty-year struggle. In *From Redlining to Reinvestment* (ed. G. Squires), pp. 133–148. Temple University Press.

[352] Poirier, Ian (2012). High-frequency trading and the flash crash: Structural weaknesses in the securities markets and proposed regulatory responses. *Hastings Business Law Journal*, **8**(2).

[353] Poitras, Laura and Greenwald, Glenn (9 June 2013). NSA whistleblower Edward Snowden: 'I don't want to live in a society that does these sort of things' – video. https://www.theguardian.com/world/video/2013/jun/09/nsa-whistleblower-edward-snowden-interview-video. The Guardian US, Online, accessed 28 February 2020.

[354] Porter, Jon (6 February 2020). Facebook and Linkedin are latest to demand clearview stop scraping images for facial recognition tech. https://www.theverge.com/2020/2/6/21126063/facebook-clearview-ai-image-scraping-facial-recognition-database-terms-of-service-twitter-youtube. theverge.com, Online, accessed 1 October 2020.

[355] Powell, Betsy (13 April 2020). How Toronto police used controversial facial recognition technology to solve the senseless murder of an innocent man. https://www.thestar.com/news/gta/2020/04/13/how-toronto-police-used-controversial-facial-recognition-technology-to-solve-the-senseless-murder-of-an-innocent-man.html. The Star, Online, accessed 1 June 2020.

[356] Praet, Stiene, Aelst, Peter Van, and Martens, David (2018). I like, therefore I am: predictive modeling to gain insights in political preference in a multi-party system. Research Report 2018-014, University of Antwerp, Faculty of Applied Economics.

[357] Praet, Stiene and Martens, David (2020). Efficient parcel delivery by predicting customers' locations. *Decision Science*, **51**(5), 1202–1231.

[358] Preter, Wim De (29 May 2019). Burgemeester beboet voor misbruik per-soonsgegevens in campagne. https://www.tijd.be/politiek-economie/belgie/verkiezingen/burgemeester-beboet-voor-misbruik-persoonsgegevens-in-campagne/10131666.html. De Tijd, Online, accessed 7 March 2019.

[359] Price, Rob (24 March 2016). Microsoft is deleting its ai chatbot's incredibly racist tweets. https://www.businessinsider.com/microsoft-deletes-racist-genocidal-tweets-from-ai-chatbot-tay-2016-3. Online, accessed 25 May 2021.

[360] Prinz, Florian, Schlange, Thomas, and Asadullah, Khusru (2011, September). Believe it or not: how much can we rely on published data on potential drug targets? *Nature Reviews Drug Discovery*, **10**(9), 712–712.

[361] Profis, Sharon (22 May 2014). Do wristband heart trackers actually work? a checkup. https://www.cnet.com/news/how-accurate-are-wristband-heart-rate-monitors/. cnet, Online, accessed 28 February 2020.

[362] Provost, Foster, Dalessandro, Brian, Hook, Rod, Zhang, Xiaohan, and Murray, Alan (2009). Audience selection for on-line brand advertising: privacy-friendly social network targeting. In *Proceedings of the 15th ACM SIGKDD International Conference on Knowledge Discovery and Data Mining*, pp. 707–716. ACM.

[363] Provost, Foster and Fawcett, Tom (2013). Data science and its relationship to big data and data-driven decision making. *Big Data*, **1**(1).

[364] Provost, Foster and Fawcett, Tom (2013). *Data Science for Business: What You Need to Know about Data Mining and Data-Analytic Thinking* (1st edn). O'Reilly Media, Inc.

[365] Provost, Foster, Martens, David, and Murray, Alan (2015). Finding similar mobile consumers with a privacy-friendly geo-social design. *Information Systems Research*, **26**(2), 243–472.

[366] Pu, Jiameng, Mangaokar, Neal, Kelly, Lauren, Bhattacharya, Parantapa, Sundaram, Kavya, Javed, Mobin, Wang, Bolun, and Viswanath, Bimal (2021). Deepfake videos in the wild: Analysis and detection.

[367] Pulitzer Prize Board (1989). 1989 Pulitzer Prizes – journalism. https://www.pulitzer.org/prize-winners-by-year/1989. Online, accessed 21 January 2021.

[368] Quill (11 April 2011). Gone but not forgotten: Ethics guidelines from phil record. https://www.quillmag.com/2011/04/05/gone-but-not-forgotten-ethics-guidelines-from-phil-record/. Online, accessed 25 May 2021.

[369] Raab, Charles (2012, 01). *The Meaning of "Accountability" in the Information Privacy Context*, pp. 15–32.

[370] Ramon, Yanou, Martens, David, Evgeniou, Theodoros, and Praet, Stiene (2021). Can metafeatures help improve explanations of prediction models when using behavioral and textual data? *Machine Learning*, **163**.

[371] Randazzo, Ryan (11 December 2018). A slashed tire, a pointed gun, bullies on the road: Why do waymo self-driving vans get so much hate? https://eu.azcentral.com/story/money/business/tech/2018/12/11/waymo-self-driving-vehicles-face-harassment-road-rage-phoenix-area/2198220002/. Online, accessed 8 March 2021.

[372] Reddit (2021). Deep cage. https://www.reddit.com/r/deepcage/. Online, accessed 25 May 2021.

[373] Ribeiro, Marco Tulio, Singh, Sameer, and Guestrin, Carlos (2016). 'Why should I trust you?': Explaining the predictions of any classifier. In *Proceedings of the 22nd ACM SIGKDD International Conference on Knowledge Discovery and Data Mining*, KDD '16, New York, NY, USA, pp. 1135–1144. ACM.

[374] Richards, Neil (2015). *Intellectual Privacy: Rethinking Civil Liberties in the Digital Age*. Oxford University Press.

[375] Ripley, Brian D. and Hjort, N. L. (1995). *Pattern Recognition and Neural Networks* (1st edn). Cambridge University Press, New York, NY, USA.

[376] Rivest, R. L., Adleman, L., and Dertouzos, M. L. (1978). On data banks and privacy homomorphisms. *Foundations of Secure Computation, Academia Press*, 169–179.

[377] Rivest, Ronald L., Shamir, Adi, and Adleman, Leonard M. (1983, September). Cryptographic communications system and method. Technical Report US Patent number 4.405.829.

[378] Robert K. Nelson, LaDale Winling, Richard Marciano Nathan Connolly et al. (2021). Mapping inequality. https://dsl.richmond.edu/panorama/redlining/#loc= 11/40.794/-74.143&city=manhattan-ny. Online, accessed 21 January 2021.

[379] Romero, Simon (31 December 2018). Wielding rocks and knives, arizonans attack self-driving cars. https://www.nytimes.com/2018/12/31/us/waymo-self-driving-cars-arizona-attacks.html. New York Times, Online, accessed 8 March 2021.

[380] Rosenberg, Matthew, Confessore, Nicholas, and Cadwalladr, Carole (17 March 2018). How Trump consultants exploited the Facebook data of millions. https://www.nytimes.com/2018/03/17/us/politics/cambridge-analytica-trump-campaign.html. Online; accessed 7 March 2019.

[381] Sabine Hossenfelder (2 August 2019). Quantum supremacy is coming. it won't change the world. https://www.theguardian.com/technology/2019/aug/02/ quantum-supremacy-computers. The Guardian, Online, accessed 4 August 2019.

[382] Samarati, Pierangela and Sweeney, Latanya (1998). Protecting privacy when disclosing information: k-anonymity and its enforcement through generalization and suppression. Technical report.

[383] Samek, Wojciech, Montavon, Gregoire, Lapuschkin, Sebastian, Anders, Christopher J., and Muller, Klaus-Robert (2021, March). Explaining deep neural networks and beyond: A review of methods and applications. *Proceedings of the IEEE*, **109**(3), 247–278.

[384] Sander, David E. and Chen, Brian X. (26 September 2014). Signaling post-Snowden era, new iPhone locks out N.S.A. https://www.nytimes.com/2014/09/27/technology/ iphone-locks-out-the-nsa-signaling-a-post-snowden-era-.html. New York Times, Online, accessed 28 October 2019.

[385] Sandler, Mark, Howard, Andrew, Zhu, Menglong, Zhmoginov, Andrey, and Chen, Liang-Chieh (2018, June). Mobilenetv2: Inverted residuals and linear bottlenecks. In *Proceedings of the IEEE Conference on Computer Vision and Pattern Recognition (CVPR)*.

[386] Sanger, David E. and Chen, Brian X. (27 Sepember 2014). Signaling post-snowden era, new iphone locks out nsa. https://www.nytimes.com/2014/09/27/technology/ iphone-locks-out-the-nsa-signaling-a-post-snowden-era-.html. New York Times, Online, accessed 1 May 2020.

[387] Sanger, David E. and Frenkel, Sheera (4 September 2018). 'Five eyes' nations quietly demand government access to encrypted data. https://www.nytimes.com/2018/09/ 04/us/politics/government-access-encrypted-data.html. New York Times, Online, accessed 28 February 2020.

[388] Schmidt, Ulf, Frewer, Andreas, and Sprumont, Dominique (2020). *Ethical Research: The Declaration of Helsinki, and the Past, Present and Future of Human Experimentation*. Oxford University Press.

[389] Schor, Peter (1994). Algorithms for quantum computation: Discrete logarithms and factoring. In *Proceedings of the 35th Annual Symposium on the Foundations of Computer Science*, pp. 124–134. IEEE Computer Society.

[390] Schwab, Klaus and Zahidi, Saadia (2020). The future of jobs report 2020. http://www3.weforum.org/docs/WEF_Future_of_Jobs_2020.pdf. Online, accessed 8 March 2021.

[391] Schwartz, Oscar (25 November 2019). In 2016, Microsoft's racist chatbot revealed the dangers of online conversation. https://spectrum.ieee.org/tech-talk/artificial-intelligence/machine-learning/in-2016-microsofts-racist-chatbot-revealed-the-dangers-of-online-conversation. Online, accessed 25 May 2021.

[392] Scoot Galloway (25 February 2016). Scott Galloway on Apple vs FBI: Sorry not sorry. https://www.l2inc.com/daily-insights/scott-galloway-on-apple-vs-fbi-sorry-not-sorry. Gartner L2, Online, accessed 13 September 2019.

[393] Shearer, Colin (2000). The Crisp-DM model: The new blueprint for data mining. *Journal of Data Warehousing*, 5(4).

[394] Sheehan, Matt (19 December 2018). How Google took on China and lost. https://www.technologyreview.com/2018/12/19/138307/how-google-took-on-china-and-lost/. MIT Technology Review, Online, accessed 1 October 2020.

[395] Shmueli, Galit (2010). To explain or to predict? *Statistical Science*, 25(3), 289–310.

[396] Shmueli, Galit (2017). Research dilemmas with behavioral big data. *Big Data*, 5(2), 98–119.

[397] Shmueli, Galit (2020). 'Improving' prediction of human behavior using behavior modification.

[398] Shmueli, Galit, Bruce, Peter C., and Patel, Nitin R. (2016). *Data Mining for Business Analytics: Concepts, Techniques, and Applications with XLMiner* (3rd edn). Wiley Publishing.

[399] Sigh, Simon (2000). *The Code Book: The Science of Secrecy from Ancient Egypt to Quantum Cryptography*. Anchor.

[400] Simon, Phil (March, 2014). Potholes and big data: Crowdsourcing our way to better government. https://www.wired.com/insights/2014/03/potholes-big-data-crowdsourcing-way-better-government/. Wired, Online, accessed 11 November 2019.

[401] Simonite, Tom (10 July 2019). Who's listening when you talk to your google assistant? https://www.wired.com/story/whos-listening-talk-google-assistant/. Wired, Online, accessed 28 February 2020.

[402] Simonite, Tom (1 November 2018). When it comes to gorillas, Google Photos remains blind. https://www.wired.com/story/when-it-comes-to-gorillas-google-photos-remains-blind/. Wired, Online, accessed 1 May 2020.

[403] Singer, Natasha (17 May 2014). Never forgetting a face. https://www.nytimes.com/2014/05/18/technology/never-forgetting-a-face.html. New York Times, Online, accessed 1 June 2020.

[404] Smith, Alexander (23 November 2015). #brussellock cat tweets go viral during terrorism raids. https://www.nbcnews.com/storyline/paris-terror-attacks/brusselslockdown-cat-tweets-go-viral-during-terrorism-raids-n468021. NBC News, Online, accessed 28 February 2020.

[405] Smith, Elise and Master, Zubin (2014, 12). *Ethical Practice of Research Involving Humans*, pp. 1–11.

[406] Snow, Jacob (26 July 2018). Amazon's face recognition falsely matched 28 members of Congress with mugshots. https://www.aclu.org/blog/privacy-technology/surveil-

lance-technologies/amazons-face-recognition-falsely-matched-28. ACLU, Online, accessed 1 June 2020.

[407] Snowden, Edward (21 May 2015). Just days left to kill mass surveillance under section 215 of the patriot act. https://www.reddit.com/r/IAmA/comments/36ru89/just_days_left_to_kill_mass_surveillance_under/crglgh2/. Reddit, Online, accessed 1 May 2020.

[408] Social Security Administration (2011). Social security number randomization. https://www.ssa.gov/employer/randomization.html. Online, accessed 25 May 2021.

[409] Sokol, Kacper and Flach, Peter (2020, Jan). Explainability fact sheets. *Proceedings of the 2020 Conference on Fairness, Accountability, and Transparency.*

[410] South, Jeff (7 August 2018). More than 1,000 U.S. news sites are still unavailable in europe, two months after GDPR took effect. https://www.niemanlab.org/2018/08/more-than-1000-u-s-news-sites-are-still-unavailable-in-europe-two-months-after-gdpr-took-effect/. Nieman Lab, Online, accessed 1 October 2020.

[411] Sprenger, Polly (26 January 1999). Sun on privacy: 'get over it'. https://www.wired.com/1999/01/sun-on-privacy-get-over-it/. Online, accessed 25 May 2021.

[412] Stanford Encyclopedia of Philosophy (15 June 2018). Aristotle's ethics. https://plato.stanford.edu/entries/aristotle-ethics/. Online, accessed 8 March 2021.

[413] Stapel, Diederik (2012). *Ontsporing.* Prometheus.

[414] Statt, Nick and Vincent, James (7 June 2018). Google pledges not to develop ai weapons, but says it will still work with the military. https://www.theverge.com/2018/6/7/17439310/google-ai-ethics-principles-warfare-weapons-military-project-maven. The Verge, Online, accessed 1 May 2020.

[415] Stefan Wojcik and Adam Hughes (24 April 2019). Sizing up twitter users. https://www.pewinternet.org/2019/04/24/sizing-up-twitter-users/. Pew Research, Online, accessed 10 October 2019.

[416] Strubell, Emma, Ganesh, Ananya, and McCallum, Andrew (2019). Energy and policy considerations for deep learning in NLP. *CoRR*, abs/1906.02243.

[417] Subrahmanyam, Avanidhar (2013). Algorithmic trading, the flash crash, and coordinated circuit breakers. *Borsa Istanbul Review*, **13**(3), 4–9.

[418] Suciu, Peter (11 December 2020). Deepfake star wars videos portent ways the technology could be employed for good and bad. https://www.forbes.com/sites/petersuciu/2020/12/11/deepfake-star-wars-videos-portent-ways-the-technology-could-be-employed-for-good-and-bad/?sh=1339b1176437. Online, accessed 25 May 2021.

[419] Supp, Catherine (30 August 2019). Fraudsters used ai to mimic ceo's voice in unusual cybercrime case. https://www.wsj.com/articles/fraudsters-use-ai-to-mimic-ceos-voice-in-unusual-cybercrime-case-11567157402. Online, accessed 8 March 2021.

[420] Sweeney, Latanya (1997). Datafly: A system for providing anonymity in medical data. In *Proceedings of the IFIP TC11 WG11.3 Eleventh International Conference on Database Securty XI: Status and Prospects*, GBR, pp. 356–381. Chapman & Hall, Ltd.

[421] Sweeney, Latanya (1997). Weaving technology and policy together to maintain confidentiality. *The Journal of Law, Medicine & Ethics*, **25**(2-3), 98–110.

[422] Sweeney, Latanya (2000). Simple demographics often identify people uniquely.

[423] Tan, Pang-Ning, Steinbach, Michael, and Kumar, Vipin (2005, May). *Introduction to Data Mining.* Addison Wesley.

[424] Tanfani, Joseph (27 March 2018). Race to unlock San Bernardino shooter's iPhone was delayed by poor FBI communication, report finds. https://www.latimes.com/

politics/la-na-pol-fbi-iphone-san-bernardino-20180327-story.html. Los Angeles Times, Online, accessed 28 February 2020.

[425] Terjesen, Siri (5 May 2021). Why Facebook created its own 'Supreme Court' for judging content – 6 questions answered. https://theconversation.com/why-facebook-created-its-own-supreme-court-for-judging-content-6-questions-answered-160349. Online, accessed 25 May 2021.

[426] The New York Times (10 April 2018). Mark Zuckerberg testimony: Senators question Facebook's commitment to privacy. https://www.nytimes.com/2018/04/10/us/politics/mark-zuckerberg-testimony.html. Online, accessed 25 May 2021.

[427] The Retraction Watch Database (2020). Retractions of Diederik A. Stapel. http://retractiondatabase.org/RetractionSearch.aspx?AspxAutoDetectCookieSupport=1#?AspxAutoDetectCookieSupport%3d1%26auth%3dStapel%252c%2bDiederik%2bA. Online, accessed 1 October 2020.

[428] The World Bank (2020). Individuals using the internet (% of population) - united states, china. https://data.worldbank.org/indicator/IT.NET.USER.ZS?locations=US-CN. The World Bank, Online, accessed October 1, 2020.

[429] The World Bank (2020). Population, total – United States, China. https://data.worldbank.org/indicator/SP.POP.TOTL?locations=US-CN. The World Bank, Online, accessed 1 October 2020.

[430] Thomas, S.B. and Quinn, S.C. (1991). Public health then and now: The Tuskegee Syphilis Study, 1932 to 1972: Implications for HIV education and AIDS risk education programs in the black community. *American Journal of Public Health*, **81**(11), 1498–1504.

[431] Thomson, Judith Jarvis (1976). Killing, letting die, and the trolley problem. *The Monist*, **59**(2), 204–217.

[432] Tobback, Ellen and Martens, David (2019, 05). Retail credit scoring using fine-grained payment data. *Journal of the Royal Statistical Society: Series A (Statistics in Society)*, **182**.

[433] Tom Lauricella, Kara Scannell and Strasburg, Jenny (2 October 2010). How a trading algorithm went awry. https://www.wsj.com/articles/SB10001424052748704029304575526390131916792. Online, accessed 25 May 2021.

[434] Tranquillus, Gaius Suetonius (121). *About the Life of the Caesars*.

[435] Tressler, Colleen (27 August 2018). Selling your car? clear your personal data first. https://www.consumer.ftc.gov/blog/2018/08/selling-your-car-clear-your-personal-data-first. Federal Trade Commission – Consumer Information, Online, accessed 19 July 2019.

[436] Tsvetkova, M., García-Gavilanes, Ruth, Floridi, L., and Yasseri, T. (2017). Even good bots fight: The case of wikipedia. *PLoS ONE*, **12**(2).

[437] United States Census Bureau (2019). Chanlder City, Arizona. https://www.census.gov/quickfacts/chandlercityarizona. Online, accessed 8 March 2021.

[438] Univers Redactie (31 October 2011). Het leven van diederik stapel, een overzicht. https://universonline.nl/nieuws/2011/10/31/het-leven-van-diederik-stapel-een-overzicht/. Online, accessed 25 May 2021.

[439] U.S. Bureau of Labor Statistics (2019). Cashiers. https://www.bls.gov/ooh/sales/cashiers.htm. Online, accessed 8 March 2021.

[440] U.S. Bureau of Labor Statistics (2019). Employment projections: Occupations with the largest job declines. https://www.bls.gov/emp/tables/occupations-largest-job-declines.htm. Online, accessed 8 March 2021.

[441] U.S. Bureau of Labor Statistics (2019). Employment projections: Occupations with the largest job growth. https://www.bls.gov/emp/tables/occupations-most-job-growth.htm. Online, accessed 8 March 2021.

[442] U.S. Bureau of Labor Statistics (2020). Employment projections: 2019-2029 summary. https://www.bls.gov/news.release/ecopro.nr0.htm. Online, accessed 8 March 2021.

[443] U.S. Census. U.s. census. https://www.census.gov/programs-surveys/decennial-census/2020-census/about.htmlhttps://www.census.gov/programs-surveys/economic-census/guidance/data-uses.html. U.S. Census Bureau, Online, accessed 1 October 2020.

[444] U.S. Department of Justice (21 December 2017). The fair housing act. https://www.justice.gov/crt/fair-housing-act-1. Online, accessed 25 May 2021.

[445] Valentino-DeVries, Jennifer, Singer, Natasha, Keller, Michael H., and Krolik, Aaron (10 December 2018). Your apps know where you were last night, and they're not keeping it secret. https://www.nytimes.com/interactive/2018/12/10/business/location-data-privacy-apps.html. New York Times, Online, accessed 1 May 2020.

[446] Van Gestel, Tony and Baesens, Bart (2009). *Credit Risk Management: Basic Concepts: Financial Risk Components, Rating Analysis, Models, Economic and Regulatory Capital*. Oxford University Press.

[447] Vanhoeyveld, Jellis, Martens, David, and Peeters, Bruno (2020). Customs fraud detection. *Pattern Analysis and Applications*, **23**(3), 1457–1477.

[448] Varian, Hal R. (1997). Versioning information goods. https://www-inst.cs.berkeley.edu/~eecsba1/sp97/reports/eecsba1b/Final/version.pdf. Research Report.

[449] Verfaellie, Mieke and McGwin, Jenna (December 2011). The case of Diederik Stapel. MiekeVerfaellieandJennaMcGwin. American Psychological Association, Online, accessed October 1, 2020.

[450] Verma, Sahil and Rubin, Julia (2018). Fairness definitions explained. In *Proceedings of the International Workshop on Software Fairness*, FairWare '18, New York, NY, USA, pp. 1–7. Association for Computing Machinery.

[451] Vermeire, Tom and Martens, David (2020). Explainable image classification with evidence counterfactual. Technical Report 2004.07511, arxiv.

[452] Vincent, James (5 March 2021). Tom Cruise deepfake creator says public shouldn't be worried about 'one-click fakes'. https://www.theverge.com/2021/3/5/22314980/tom-cruise-deepfake-tiktok-videos-ai-impersonator-chris-ume-miles-fisher. Online, accessed 25 May 2021.

[453] Voeller, John G. (2010). *Wiley Handbook of Science and Technology for Homeland Security, 4 Volume Set*. John Wiley & Sons Inc.

[454] Štrumbelj, Erik and Kononenko, Igor (2014, December). Explaining prediction models and individual predictions with feature contributions. *Knowl. Inf. Syst.*, **41**(3), 647–665.

[455] Wachter, S., Mittelstadt, B., and Russell, C. (2021). Bias preservation in machine learning: the legality of fairness metrics under EU non-discrimination law. West Virginia Law Review.

[456] WAFB (2 May 2016). 'Brittney Mills Act' fails to pass in LA. House Committee. https://www.wafb.com/story/31866353/brittney-mills-act-fails-to-pass-in-la-house-committee/. WAFB, Online, accessed 11 November 2019.

[457] Wagner, Claudia, Garcia, David, Jadidi, Mohsen, and Strohmaier, Markus (2015). It's a man's wikipedia? assessing gender inequality in an online encyclopedia. in The

International AAAI Conference on Web and Social Media (ICWSM2015), Oxford, May 2015.

[458] Warner, Stanley L. (1965). Randomized response: A survey technique for eliminating evasive answer bias. *Journal of the American Statistical Association*, **60**(309), 63–69.

[459] Washington Post Staff (15 December 2015). Transcript: CNN undercard GOP debate. https://www.washingtonpost.com/news/post-politics/wp/2015/12/15/transcript-cnn-undercard-gop-debate/. CNN, Online, accessed 13 September 2019.

[460] Weindling, Paul (2001). The origins of informed consent: The international scientific commission on medical war crimes, and the nuremberg code. *Bulletin of the History of Medicine*, **75**(1), 37–71.

[461] Weindling, Paul (2004). *Nazi Medicine and the Nuremberg Trials: From Medical War Crimes to Informed Consent / by Paul Julian Weindling*. Palgrave Macmillan, Houndmills, Basingstoke, Hampshire; New York.

[462] Wheeler, Nicole (1 August 2019). Publication bias is shaping our perceptions of AI. https://towardsdatascience.com/is-the-medias-reluctance-to-admit-ai-s-weaknesses-putting-us-at-risk-c355728e9028. Towards Data Science, Online, accessed 8 March 2021.

[463] Wikipedia. List of the most common passwords. https://en.wikipedia.org/wiki/List_of_the_most_common_passwords. Wikipedia, Online, accessed 19 July 2019.

[464] Wikipedia. Martin Handford. https://en.wikipedia.org/wiki/Martin_Handford. Wikipedia, Online, accessed 1 October 2020.

[465] Wikipedia (2021). Women's suffrage. https://en.wikipedia.org/wiki/Women%27s_suffrage. Online, accessed 8 March 2021.

[466] Wong, Chris (18 March 2014). FOILing NYC's Taxi Trip Data. https://chriswhong.com/open-data/foil_nyc_taxi/. Online, accessed 1 May 2020.

[467] Wood, Alexandra, Altman, Micah, Bembenek, Aaron, Bun, Mark, Gaboardi, Marco, Honaker, James, Nissim, Kobbi, OBrien, David R., Steinke, Thomas, and Vadhan, Salil (2018). Differential privacy: A primer for a non-technical audience. *Vanderbilt Journal of Entertainment & Technology Law*, **21**(1), 209–275.

[468] Wood, Molly (28 July 2014). Ok cupid plays with love in user experiments. https://www.nytimes.com/2014/07/29/technology/okcupid-publishes-findings-of-user-experiments.html. New York Times, Online, accessed 28 October 2019.

[469] Woodward, John D. (2001). *Super Bowl Surveillance: Facing Up to Biometrics*. RAND Corporation, Santa Monica, CA.

[470] World Health Association (1964). Declaration of Helsinki – Ethical principles for medical research involving human subjects. https://www.wma.net/policies-post/wma-declaration-of-helsinki-ethical-principles-for-medical-research-involving-human-subjects/. Online, accessed 28 February 2020.

[471] Wozniak, Steve (10 November 2019). Tweet. https://twitter.com/stevewoz/status/1193330241478901760. Twitter, Online, accessed 21 January 2021.

[472] Yang, Andrew (14 November 2019). Yes, robots are stealing your job. https://www.nytimes.com/2019/11/14/opinion/andrew-yang-jobs.html. Online, accessed 8 March 2021.

[473] Yao, Andrew C. (1982). Protocols for secure computations. In *Proceedings of the 23rd Annual Symposium on Foundations of Computer Science*, SFCS '82, USA, pp. 160–164. IEEE Computer Society.

[474] Yin, Juan, Cao, Yuan, Li, Yu-Huai, Liao, Sheng-Kai, Zhang, Liang, Ren, Ji-Gang, Cai, Wen-Qi, Liu, Wei-Yue, Li, Bo, Dai, Hui, Li, Guang-Bing, Lu, Qi-Ming, Gong, Yun-Hong, Xu, Yu, Li, Shuang-Lin, Li, Feng-Zhi, Yin, Ya-Yun, Jiang, Zi-Qing, Li, Ming, Jia, Jian-Jun, Ren, Ge, He, Dong, Zhou, Yi-Lin, Zhang, Xiao-Xiang, Wang, Na, Chang, Xiang, Zhu, Zhen-Cai, Liu, Nai-Le, Chen, Yu-Ao, Lu, Chao-Yang, Shu, Rong, Peng, Cheng-Zhi, Wang, Jian-Yu, and Pan, Jian-Wei (2017). Satellite-based entanglement distribution over 1200 kilometers. *Science*, **356**(6343), 1140–1144.

[475] Yoori Hwang, Ji Youn Ryu and Jeong, Se-Hoon (2021). Effects of disinformation using deepfake: The protective effect of media literacy education. *Cyberpsychology, Behavior, and Social Networking*, **24**(3), 188–193.

[476] Zach Epstein (25 February 2016). San Bernandino iPhone interview Tim Cook. https://bgr.com/2016/02/25/san-bernardino-iphone-interview-tim-cook/. BGR, Online, accessed 7 October 2019.

[477] Zafar, Muhammad Bilal, Valera, Isabel, Gomez Rodriguez, Manuel, and Gummadi, Krishna P. (2017). Fairness beyond disparate treatment and disparate impact: Learning classification without disparate mistreatment. WWW '17, Republic and Canton of Geneva, CHE, pp. 1171–1180. International World Wide Web Conferences Steering Committee.

[478] Zeller, Tom Jr. (22 August 2006). AOL executive quits after posting of search data – technology – International Herald Tribune. https://www.nytimes.com/2006/08/22/technology/22iht-aol.2558731.html. New York Times, Online, accessed 1 May 2020.

[479] Zemel, Rich, Wu, Yu, Swersky, Kevin, Pitassi, Toni, and Dwork, Cynthia (2013, 17–19 June). Learning fair representations. Volume 28 of *Proceedings of Machine Learning Research*, Atlanta, Georgia, USA, pp. 325–333. PMLR.

[480] Zhang, Brian Hu, Lemoine, Blake, and Mitchell, Margaret (2018). Mitigating unwanted biases with adversarial learning. In *Proceedings of the 2018 AAAI/ACM Conference on AI, Ethics, and Society*, AIES '18, New York, NY, USA, pp. 335–340. Association for Computing Machinery.

[481] Zuboff, Shoshana (2018). *The Age of Surveillance Capitalism: The Fight for a Human Future at the New Frontier of Power* (1st edn).

[482] Zuckerberg, Mark (5 May 2021). A blueprint for content governance and enforcement. https://www.facebook.com/notes/751449002072082/. Online, accessed 25 May 2021.

Index